Oils and Fats Authentication

Chemistry and Technology of Oils and Fats

Series Editor: R.J. Hamilton

A series which presents the current state of the art in chosen areas of oils and fats chemistry, including its relevance to the food and pharmaceutical industries. Written at professional and reference level, it is directed at chemists and technologists working in oils and fats processing, the food industry, the oleochemicals industry and the pharmaceutical industry, at analytical chemists and quality assurance personnel, and at lipid chemists in academic research laboratories. Each volume in the series provides an accessible source of information on the science and technology of a particular area.

Titles in the series:

Spectral Properties of Lipids
Edited by R.J. Hamilton and J. Cast

Lipid Synthesis and Manufacture
Edited by F.D. Gunstone

Edible Oil Processing
Edited by R.J. Hamilton and W. Hamm

Oleochemical Manufacture and Applications
Edited by F.D. Gunstone and R.J. Hamilton

Oils and Fats Authentication
Edited by M. Jee

Vegetable Oils in Food Technology
Edited by F.D. Gunstone

Oils and Fats Authentication

Edited by

MICHAEL JEE
Head of Lipids Section
Reading Scientific Services Ltd
Reading, UK

Blackwell
Publishing

CRC Press

Editorial Offices:
Osney Mead, Oxford OX2 0EL, UK
Tel: +44 (0)1865 206206
108 Cowley Road, Oxford OX4 1JF, UK
Tel: +44 (0)1865 791100
Blackwell Munksgaard, Nørre Søgade 35, PO Box 2148, Copenhagen, DK-1016, Denmark
Tel: +45 77 33 33 33
Blackwell Publishing Asia, 54 University Street, Carlton, Victoria 3053, Australia
Tel: +61 (0)3 9347 0300
Blackwell Verlag, Kurfürstendamm 57, 10707 Berlin, Germany
Tel: +49 (0)30 32 79 060
Blackwell Publishing, 10 rue Casimir Delavigne, 75006 Paris, France
Tel: +33 1 53 10 33 10

ISBN 1-84127-330-9

Published in the USA and Canada (only) by
CRC Press LLC
2000 Corporate Blvd., N.W.
Boca Raton, FL 33431, USA
Orders from the USA and Canada (only) to
CRC Press LLC

USA and Canada only:
ISBN 0-8493-2815-2

First published 2002

British Library
Cataloguing-in-Publication Data:
A catalogue record for this title is available from the British Library

Library of Congress
Cataloging-in-Publication Data:
A catalog record for this book is available from the Library of Congress

Set in 10½/12pt Times
by Thomson Press (India) Ltd

Preface

Quality assessment and the need for authentication are increasingly important features of the food and personal care industries. Yet, although there have been articles in journals and chapters in books which have described techniques of authentication, a book devoted entirely to this subject has not previously been available.

This book provides an overview of the methods relevant to authentication of the major oils and fats. The first chapter presents an introduction to the techniques used to evaluate the purity of oils, describing the problems which can arise—particularly in relation to products labelled 'organic', 'unrefined' or 'GM-free'. The approaches that may be used in addressing these problems are described and likely developments are considered.

The only oils in commerce that do have a legal definition backed by officially sanctioned methods of analysis are the various grades of olive oil. The second chapter discusses the production of these oils and its relationship to the various grades, and the officially sanctioned methods of analysis are described by an acknowledged expert in this field.

Unlike olive oil, the analysis of cocoa butter is not governed by legal definitions. However, the legal definition of chocolate is specific in relation to whether cocoa butter is present alone or as the major vegetable fat, with strict limits on the presence of other vegetable fats in the product. Cocoa butter is also one of the few fats for which artificially manufactured substitutes of similar composition have been constructed and openly marketed. Because of this, analysis of the adulteration of cocoa butter probably has a greater importance than that of any fat other than olive oil, and the approaches to this analysis are described in chapter 3.

Oils used mainly for the health benefits of specific fatty acids are always a potential source of adulteration. Chapter 4 investigates three types of these oils: fish oils, evening primrose oil and borage oil (once used as an adulterant of evening primrose oil, but now an acknowledged oil in its own right). The first two are reasonably well defined, but fish oils—because of the range of sources and compositions—still present a problem for analysts.

Chapter 5 describes milk and other animal fats. Expensive bovine milk fats have often been adulterated in the past and, with the increased marketing of milk products from other species, this is an area requiring investigation. One might think that carcass animal fats are not likely to be adulterated, as they are often cheaper than vegetable alternatives. While this is a reasonable initial assumption, there is a considerable market for foods and products which, for religious reasons, do not contain pig fat. This area of analysis has not, to my knowledge, been summarised previously. This chapter also describes the problems encountered in relation to vegetarian or vegan products. After the initial writing of the chapter, BBC World News reported pressure in India for cosmetics to be tested and labelled according to whether only vegetarian ingredients were used in their manufacture. The difficulty and, in many cases, impossibility of carrying this out is described here.

Chapters 6 and 7 focus on techniques used in checking for authenticity. The most useful components for detecting sophisticated adulteration are the minor components. These analyses often produce a mass of data in which a pattern is difficult to detect with an untrained eye. Chemometrics can be utilised to investigate trends and patterns and thus detect non-standard oils which might otherwise be missed.

There are many reasons for wanting to authenticate oil. Suitability for purpose, taste or religious or moral requirements are important but, for commercial organisations, the primary reasons are due diligence and legal requirements. The final chapter, written by one well versed in the arguments, describes the legal issues raised by authenticity and adulteration of oils.

I would like to thank the contributors for their work on this volume. Our publisher, Dr Graeme MacKintosh, was keen to produce a book that was as up-to-date as possible, and so a certain amount of encouragement (or gentle prodding) was required. I hope that the resultant product will serve as a useful source of reference to this important area.

Michael Jee

Contributors

Dr Ramón Aparicio	Instituto de la Grasa, Avenida Padre Garcia Tajero, 4, 41012 Sevilla, Spain
Mr Ramón Aparicio-Ruiz	Muelle de Heredia, 20 29001 Malaga, Spain
Professor Giorgio Bianchi	Istituto Sperimentale per la Elaiotecnica, Contrada Fonte Umano 37, 65013 Città Sant'Angelo (Pescara), Italy
Mr Colin Crews	Central Science Laboratory, Sand Hutton, York YO41 1LZ, UK
Professor N.A. Michael Eskin	Department of Human Nutritional Sciences, Faculty of Human Ecology, University of Manitoba, Winnipeg, Manitoba, Canada, RET 2N2
Dr Michael H. Gordon	School of Food Biosciences, The University of Reading, Whiteknights, PO Box 226, Reading RG6 6AP, UK
Dr Michael Jee	Reading Scientific Services Ltd, Units 25-27, Robert Cort Estate, Britten Road, Reading RG2 0AQ, UK
Ms Catriona Stewart	Food Labelling and Standards Division, Food Standards Agency, Aviation House, 125 Kingsway, London WC2B 6NH, UK

Contents

1 Adulteration and authentication of oils and fats: an overview

Michael Jee

1.1 Introduction

It is certain that not all of the oils consumed today are completely authentic with respect to all the descriptions on the label (Grob *et al.*, 1994; Firestone, 2001; Working Party on Food Authenticity, 1996). Although there are published expected chemical compositions of the major edible oils (*Codex Alimentarius* Commission, 1997), the only oil that has a defined legal composition is olive oil (EC Council, 1991). This does not mean that adulterated oils cannot be identified, but it does mean that, in many cases, doing so is not an easy matter.

Most lay people, on mention of adulteration of oils, would probably think immediately of Spanish toxic oil syndrome (Posada *et al.*, 1996), which has also been called Spanish olive oil syndrome. Here rapeseed oil, intended for non-edible uses only, had been deliberately made non-edible by addition of aniline. Persons unknown attempted to remove the aniline by normal oil refining methods, and the product was then sold, either alone or blended, as a cheap cooking oil, sometimes as an olive oil. Presumably the sales were to rather non-discriminating customers. However, although the refining had superficially removed the aniline, it had produced harmful compounds that gave rise to serious health problems in consumers. Thus an oil was deliberately contaminated, the contamination was 'removed', the product was sold, sometimes as a completely different oil, and compounds produced by the processing caused severe neurological health problems. Yet, although this incident of 'contamination' is the best known to the general public, it is an exception to the norm. Even simple tests available a century ago would have shown that the oil was not olive. In addition, in almost all modern cases of adulterated oils, health problems are not an issue. The only other notable recent exception to this was the 1998 adulteration of mustard oil on the Indian subcontinent with poisonous argemone oil. Although action was taken, it was reported (*Kathmandu Post*, 2001) that in Nepal in 2000 66% of rapeseed oil was still contaminated in this way.

In most, though not all, cases there is no way that an average or sophisticated consumer could ever know, without scientific testing, that a non-authentic oil was not what it claimed to be. Indeed, at least with 'bulk' food oils, such as corn, sunflower etc., a low level (e.g. 1%) of another oil being present would often be accepted in a product as being probably due to accidental mixing in the refining plant. This is not merely because these levels are often difficult or impossible to determine; from a practical point of view it can make no difference to the product, and it could be of no conceivable economic benefit to deliberately adulterate at this level.

Nowadays, as will be seen in this book, adulteration—particularly of expensive oils—is often very sophisticated. Thus authentication of an oil is necessarily also very sophisticated and usually involves a number of different approaches. This was not always so.

1.2 Early adulteration and its detection

Although adulteration of oils is much discussed today, the authenticity of oils and fats is not a problem that has been confined to recent years. Before the nineteenth century, many of the products on sale were not what they seemed or claimed to be. This undoubtedly includes oils. However, the methods of proving the authenticity of oils were somewhat limited. Olive oil could be tested to differentiate it from lard oil or a mixture with lard oil by cooling or by the reaction of a solution of mercury in nitric acid on the sample (Noah, 1844; Mitchell, 1848). This was of particular importance to Jewish consumers. Specific gravity was also often used (Mitchell, 1848). Tallow could be tested for greases by examination and smelling of the released fatty acids (Anon, 1856). In England a select committee (Postgate, 1885) was told that, amongst the many non-oil items, cod liver oil was often diluted with bland oils, lard with mutton fat and butter with lard. These statements were later confirmed by others (Anon 1856; Hassall 1861), while it was also known that cod liver oil was often partially substituted by other fish oils (Anon 1856). The last of these cases might still pose a problem, both in occurrence and detection, even today.

In the UK at least, the Adulteration Act of 1860, the result of the deliberations of the above committee, was the beginning of a more scientific approach to authentication of fats and oils. However, it was still being stated after the turn of the century (Sloane 1907) that, in the USA, butter was being adulterated by oleomargarine and lard, and cream by cottonseed oil and other fats. Indeed the USA equivalent of the UK Adulteration Act, the 1938 Federal Food, Drug and Cosmetic Act, was only passed after a series of even later cases of adulteration: coconut and cottonseed replacing cocoa butter and milk-fat (1922), peanut oil in olive oil (1923), lard contaminating butter (1926) and sesame oil in olive oil used in tinned sardines (1936) (Kurtzweil, 1999).

1.3 Introduction of more systematic methods of detecting adulteration

Although there were many impure oils and fats in the marketplace in 1907, the methods for detecting them were in many cases becoming available. The importance of oils and fats in the economy, together with the expansion of possible uses, meant that chemists were amassing large quantities of data on the properties of oils and fats, both edible and non-edible. A textbook which includes many of these determined properties and developments was Lewkowitsch (1895, 1904), much later replaced by Hilditch and Williams (1964).

By 1904, Lewkowitsch contained over 1000 pages on the processing, properties and methods of analysis of over 220 different oils from acorn oil, through purging nut and sod oils to yellow acacia oil. A typical list of some of the values recorded, in this case for tallow, is shown in Figure 1.1. In many cases likely adulterants, and tests which might be used to detect them, are given. Thus possible contaminants of olive oils are listed as *Arachis* (peanut), sesame, cotton seed, rape, castor, physic-nut (curcas), lard, drying oils, hydrocarbons and fish oils. Iodine values are recommended as helpful for checking six of the above, with additional tests also listed. Castor oil contamination was detected by specific gravity, acetyl value and solubility in solvents. The simplest test for the latter oil might have been to consume the oil and observe the unfortunate results, but this is not suggested.

Many of the tests described involve physical properties such as refractive index, viscosity or melting point of the fat, of the fatty acids or of the lead salts of the fatty acids. However, there were also many chemical tests such as Reichert, Polenske, iodine, saponification and acetyl values. These all gave information as to the composition of the fat, some information as to fatty acid composition, others as to other non-glyceride components of the fat. Thus the iodine value is a measure of unsaturated fatty acids in the fat, now obtainable in more detail from a fatty acid profile. Similarly the Reichert value is a measure of volatile fatty acids soluble in water. For most purposes this means butyric acid, and so the modern equivalent is the determination of butyric acid in the oil. The modern method for milk-fat analysis is thus carrying out the analysis in a similar way to the Reichert determination, but uses a technique that is less dependent on the exact conditions of the analysis and is thus less likely to be subject to operator error. The Reichert value could be useful, in theory, even if milk fat was not present. Lewkowitsch notes that some other oils do give high values. Porpoise jaw oil has a value almost twice that of milk fat, while some other oils also have significant values. It is unlikely that one would have come across much porpoise jaw oil even in 1904, and even less likely today.

Some of the tests involved relatively simple colour reactions such as the Baudouin reaction for sesame oil, and the Halphen test for cottonseed oil. In both cases a compound characteristic to an oil is used to determine the presence of the oil. Here again the test detected a component that today would be detected and quantified by gas chromatography (GC) or high performance liquid chromatography (HPLC). It was even possible to determine the presence of cholesterol or phytosterols, although, after separation, the identification as to which type was present depended on microscopic examination and fractional

Physical and Chemical Constants of Beef Tallow

Specific Gravity			Solidifying Point		Melting Point		Saponification Value		Iodine Value		Reichert Value	Hehner Value		Thermal Test: Heat of Bromination		Refractive Index	
At °C		Observer	°C	Observer	°C	Observer	Mgrms KOH	Observer	Per cent	Observer	c.c. 1/10 norm KOH	Per cent	Observer	°C	Observer	At 60° C	Observer
15	0.943–0.952	Dieterich	35–27 Temperature rises a few degrees without, however, remaining constant	Rüdorff	45–46; 42–43, not below 40	Wolkenhaar	195.7–200	Filsinger	40	Hübl	0.25	95.6	Bensemann	6.1[3]	Archbutt	1.4510	Thoerner
50 (water 15.5=1)	0.8950	Allen			47.6–48.5	Dieterich	193.2–198	Köttstorfer / Deering / L. Mayer	43.3–44	Wilson		95.4–96	L. Mayer	6.7–7.2[4]	"	Butyro-refractometer	
98 (water 15.5=1)	0.8626	"			43.5–45	Rüdorff			35.4–36.4	Filsinger							
									35.6–38.9	Dieterich							
									45.2[1]	Wallenstein and Finck							
100 (water 15=1)	0.860–0.861	Wolkenhaar							38.3[2]	"						At 40° C	Observer
100 (water 15=1)	0.860	Koenigs							38–40	Thoerner						49	Mansfeld
									41–47.5	Lewkowitsch							

Physical and Chemical Constants of the Mixed Fatty Acids of Beef Tallow

Specific Gravity		Solidifying Point		Melting Point		Neutralisation Value		Mean Molecular Weight		Iodine Value		Refractive Index	
At 100°C (water at 100=1)	Observer	°C	Observer	°C	Observer	Mgrms KOH	Observer		Observer	Per cent	Observer	At 60°C	Observer
0.8698	Archbutt	Titer Test		43–44 and 46–47	Bensemann	197.2	Lewkowitsch	270–285	A. Wright	41.3	Williams	1.4375	Thoerner
		43.5–45	Dalican	45	Hübl	201.6	Thoerner	284.5	Lewkowitsch	25.9–32.8	Morawski and Demski	Oleo-refractometer	
		44.5	De Schepper and Geitel							Iodine Value of the Liquid Fatty Acids		"Degrees"	Observer
		43	Hübl							92.4[1]	Wallenstein and Finck	−40	Jean
		Cp. p. 182	Lewkowitsch							92.2[2]	"		

[1] Australian [2] Berlin town tallow [3] American beef [4] Home-refined

Figure 1.1 Physical and chemical constants of beef tallow (from Lewkowitsch, 1904).

crystallization, followed by melting point determination. One value, the Hehner value, sometimes listed for an oil, might be considered to be a very primitive form of chemometrics. Here the sum of the insoluble fatty acids and the unsaponifiable matter for the oil was expressed as a percentage, thus combining two different tests. Many of these methods such as the Halphen test and the Fitelson test for teaseed oil are still listed in the manual of The American Oil Chemists' Society (AOCS, 1990).

The methods of fractionation listed in Lewkowitsch (1904) soon led to the separation and determination of individual fatty acids in fats. These were listed in Hilditch and Williams (1964) and the values obtained for major components were usually very similar to those later determined by GC.

1.4 Range of methods used today

Modern methods of authentication began with the development of chromatography. The first practical use of GC for any purpose was to separate the methyl esters of (short chain) fatty acids (James and Martin, 1952). The relatively straightforward determination of total fatty acid composition in effect could have replaced many of the other tests previously carried out on oils, such as iodine, Reichert and Polenske values, though these tests were still carried out for some time. Far more information was available using GC than that provided by these earlier methods of analysis. A number of very minor fatty acids, such as branched chain and odd-numbered acids, were found to be present in animal fats that were, with the exception of hexadecanoic acid, largely absent in vegetable fats (Bastlins, 1970; Wurzinger and Hensel, 1969). Sometimes trace fatty acids or glycerides such as these could be concentrated by fractional crystallization (Iverson *et al.*, 1965; Martel, 1977; von Peters and Wieske, 1966; Tan *et al.*, 1983; Synouri-Vrettakou *et al.*, 1984). Fish oils contained many fatty acids not present in other oils, though until recently, as better processing and preservative techniques became available, smell would have been just as good at detecting them. The detection of lauric acid-derived fats, such as palm kernel oil or coconut, long used in chocolate because of their similar melting characteristics to cocoa butter, became simple, at least for chocolates not containing milk fat. For those chocolates containing milk fat it was still reasonably easy, though somewhat less sensitive. Although this was quickly realized in the chocolate industry, the first publication by a regulatory source was not till 1972 (Iverson, 1972).

The methods of analysis for fatty acids (AOCS, 1990) are now one of the most frequent methods used in the analysis of fats. Further to total fatty acid determinations, it became possible, after reaction with pancreatic lipase, to determine the average fatty acid composition at the 2-position of a fat (Christie, 1986) and thus detect inter-esterification. This previously could only be detected by physical methods such as melting point. The reason that this can detect inter-esterification is that the process is carried out to affect the melting characteristics of the fat by randomizing the fatty acids in the triglyceride molecules present in the fat. In vegetable oils, the 2-position of the triglycerides largely contains only unsaturated acids, but, after inter-esterification, contains higher levels of saturated acids. In those fats, such as animal fats, where there are high levels of saturated acids at the 2-position, then the levels of unsaturated acids increase there.

It was not only fatty acids that could be analysed by GC. Triglycerides were also found to be separable, at least with respect to molecular weight. Thus triglycerides could be separated by carbon number. Later developments produced columns which also separated unsaturated triglycerides, but problems with recoveries of the more unsaturated components mean that it is only carbon number separation that is suitable for authentication purposes. The main uses of this procedure are for cocoa butters and milk fats.

GC of sterols was also found to be a very useful technique. This could be carried out either on the sterol core or on the sterol ester. Oils can contain both but, for historical reasons, and because it is a simpler procedure, the pattern and level of the sterols themselves, rather than the esters, is the more commonly used technique. Because of this there is far more data available for ranges in oils for the former (*Codex Alimentarius*, 1997; AOCS, 1997; Gutfinger and Letan, 1974; Itoh *et al.*, 1973; Rossell, 1991) than the latter. It is possible that differences between free sterols and sterol esters will become useful in checking adulteration (Youk *et al.*, 1999).

Three other GC analyses now used in authentication, largely for olive and other oils which should not be refined or solvent extracted, are the determination of waxes, aliphatic alcohols, triterpene alcohols (uvaol and erythrodiol), and stigmastadiene and other sterol-dehydration products (EEC, 1991). These analyses are used at present not to detect adulteration with other oils, but with solvent-extracted or refined oils. However, it is possible that, with solvent-extracted oils, wax, aliphatic alcohol and terpene alcohol compositions, could prove useful in differentiating or detecting different oils.

With its development, HPLC was found to be useful in many authenticity determinations, either for the same or different components to those detected by GC. Triglycerides were the most immediate application. With the exception of milk fat, now that the major components of commercial fats can be completely separated by HPLC, the patterns of components can be analysed to detect adulteration. Cocoa butter adulteration with palm fractions can be detected by the presence of excess monounsaturated and diunsaturated components from the palm fraction, while more sophisticated products may be detected by measuring dipalmitoyl-monooleoyl glycerol (POP), palmitoyl-oleoyl-stearoyl glycerol (POS) and distearoyl-monooleoyl glycerol (SOS) components. In other oils, apart from the pattern of components, the presence of any significant level of trilinolein in olive or other oils relatively low in linoleic acid can show the presence of more unsaturated oils such as in soya, sunflower (normal high linoleic type) or cottonseed at low levels (Flor *et al.*, 1993). The other main use of HPLC has been the detection of tocopherol patterns and components. These are best examined together with other analyses to identify adulteration. As with fatty acids and sterols, the ranges normally found in the major oils are listed in the *Codex Alimentarius* (1997).

One method not involving chromatography that has recently been developed is stable isotope ratio analysis. This measures the ratio of ^{13}C to ^{12}C in the oil. For plants gaining their energy from the C3 photosynthetic pathway (most oilseeds) the ratio δ^{13} is around 30, whilst for plants using the C4 pathway (maize), the value is around 15. For most purposes at present the technique is limited to detection of adulteration of maize oil. Other uses of the technique may evolve, such as the examination of the ratio within individual fatty acids or within minor components of oils such as sterols (Kelly *et al.*, 1997; Royer *et al.*, 1999a). It has recently been claimed that fatty acid data and stable carbon isotopic analysis values of bulk and individual fatty acids together can be used in distinguishing the geographical origin of olive oils (Spangenberg and Ogrinc, 2001).

One thing lacking in work on authenticity is a good database of ranges of analysis values for oils. Fatty acid composition is well covered and expected ranges of sterols and tocopherol levels are also available, at least for the major oils (*Codex Alimentarius*, 1997; FOSFA, 1994 ; AOAC, 1997). There have been some surveys of other components (Rossell, 1985; Flor *et al.*, 1993) but not all information is available in the available scientific literature. And more needs to be done. For other oils the data available are even more limited. The Leatherhead Food Research Association, UK, has carried out a survey on minor oils, but the results have not been published (J.B. Rossell, 1997, personal communication), and for some of the oils the range of samples obtainable was limited. It is to be hoped that this will be remedied in the future.

1.5 Adulteration of vegetable bulk oils (coconut, cottonseed, grapeseed, maize, palm, palm kernel, peanut, safflower, sesame and sunflower)

Rapeseed oil, soyabean oil and palm oil, being the cheapest available oils, are those most likely to be used to 'bulk-out' more expensive products. One would have thought that any oil labelled as 'rapeseed', soya or 'palm' would be 100% authentic, with the possible exception of a very small amount of contamination arising from normal processes in the refinery. This would certainly seem to be true for the first two listed oils, but not necessarily for palm oil. Figure 1.2 shows two oils. Sample B can be seen to be mainly liquid oil, yet was submitted to

Figure 1.2 Samples submitted as unrefined palm oils. A, Genuine palm oil; B, palm oil adulterated with 40–50% soyabean oil.

a UK distributor in a bottled state as 'unrefined palm oil'. Due to the unusual physical state for a product of that description, a sample was submitted for analysis. After complaint to the supplier, a further sample was received. This looked similar to the first and not only give a virtually identical analysis but also had the same packing code on the bottle. It was only at the third attempt that a satisfactory sample (A) was received. The analytical results are shown in Table 1.1 for the two adulterated samples (B(1) and B(2)), together with the good sample. The fatty acid composition of the adulterated samples shows that the oil consists of 40–50% soyabean oil. Rapeseed oil is a possibility, but does not agree with the composition of the remainder of the fatty acids. The identity of the contaminent could not have been checked by analysis of tocopherols, as both palm and soya contain γ-tocopherol. However, the absence of brassicasterol in the oil showed rapeseed oil was not present, and made soyabean oil the more likely contaminant.

The above shows that rapeseed oil can easily be detected, or eliminated, as a contaminant by sterol analysis. It is also, at least in Europe, the oil most likely to be used to 'dilute' another oil. Although low levels (as a percentage of the total sterols) have been reported in some other oils (Desbordes *et al.*, 1993), the presence of brassicasterol in an oil is good evidence of contamination in any oil from a non-*Brassica* species. It is likely that the traces reported as present in some other oils arise from contamination of the sample with rapeseed oil, or from some other *Brassica* species, or from traces of some similarly behaving non-sterol not fully separated from the sterol fraction during the work-up of the sample (Desbordes *et al.*, 1983).

Table 1.1 Analysis of unrefined palm oil samples

Fatty acid	A	B(1)	B(2)	*Codex* range for palm oil
C12:0	0.3	0.3	0.2	0–0.4
C14:0	1.0	0.5	0.5	0.5–2.0
C16:0	41.2	22.7	23.9	40.1–47.5
C16:1	0.3	0.1	0.1	0–0.6
C17:0	0.1	0.1	0.1	
C18:0	4.6	4.4	4.5	3.5–6.0
C18:1(*trans*)	<0.05	<0.05	0.1	
C18:1(*cis*)	39.6	31.0	30.8	36.0–44.0
C18:2(*trans*)	<0.05	0.1	0.1	
C18:2(*cis*)	11.7	35.4	34.3	6.5–12.0
C18:3(*trans*)	0.5	0.1	0.1	
C18:3(*cis*)	0.5	4.3	4.1	0–0.5
C20:0	0.5	0.6	0.7	0–1.0
C20:1	0.2	0.2	0.2	
C22:0	<0.05	0.2	0.3	

The best approach in checking authenticity in any of the bulk oils is by carrying out a series of analyses and comparing the results to those listed in the definitive literature (*Codex Alimentarius*, 1997; FOSFA, 1994 ; AOCS, 1997). The initial analysis should always be fatty acid composition. In those oils low in linolenic acid, such as sunflower, safflower, cottonseed and maize oil, the presence of that acid above about 1.0% is a good indication of contamination, probably with rapeseed or soya oil. In the case of oils other than grapeseed the limiting figure would be 0.5%. Low levels of lauric acid in coconut or palm kernel oils would indicate another oil was present, while levels of palmitic acid in any oil above the accepted range would probably indicate palm or a palm fraction. Levels of any other fatty acid outside the accepted range, even if only slightly, would be suspicious, and the oil should be examined further. Adulteration of sesame oil has been detected on several occasions both here and in China (Yi *et al.*, 1993) by deviations in fatty acid composition from the normal range. This can often be confirmed later by similar variations from the norm in the sterol composition.

If the fatty acid composition is within the accepted range for the oil, then other analyses should be carried out. Those used will vary with the oil. If the oil being examined is maize, then the carbon isotope ratio should be examined. This is carried out by burning the oil and determining the ratio of ^{13}C to ^{12}C. From this it is possible to detect 10% or above of an oil other than maize (Rossell, 1994).

Sterol analysis can be useful other than for detecting rapeseed oil. Accepted ranges for many oils and all major oils, given as a percentage of the total sterols, are available (*Codex Alimentarius*, 1997; FOSFA, 1994; AOCS, 1997). In all common oils the major sterol is β-sitosterol, with other sterols being present in most cases at much lower levels. Apart from brassicasterol, of particular interest with respect to the bulk oils are Δ^5- and Δ^7-stigmastenol. Δ^5-Avenasterol is present at 1–9% in palm kernel oil, but at 20–41% in the more expensive coconut oil. It is also present at 17–20% in babassu oil, though this is not normally encountered in Europe. Δ^7-Stigmastenol is present at relatively high levels in both sunflower oils (7–13% in regular high linoleic oil and 14–22% in high-oleic oils) and in safflower oils (16–23% in high linoleic oils and 13–18% in high oleic oils), but is present at relatively low levels in most other oils, including soyabean, cottonseed and peanut. The levels present of these sterols in these oils are therefore particularly useful as indicators of the purity of them.

It is claimed (Youk *et al.*, 1999) that olive, sunflower and peanut oils contain mainly esterified sterols, while soyabean and sesame oils contain mainly free sterols. This does not appear to have been utilized previously, but could be useful with mixtures of the two classes. Although it is possible to 'de-sterolize' oils, and remove characteristic sterols, this usually forms other sterols that can be detected (Biedermann and Grob, 1996; Lanuzza and Micali, 1997; Mariani and Venturini, 1997). If it is suspected that this has occurred, then the presence of other suspect components should be investigated.

Tocopherols (and tocotrienols) in an oil can also be useful, though the absolute, and in some cases the relative, levels of each can be affected by age and refining. These are usually determined by HPLC. The tocotrienols are of particular importance, as they are present in significant amounts only in corn, grapeseed, palm and palm kernel oils. The presence of γ-tocopherol in an oil that is not expected to contain significant levels, or else the presence of it in excess to that expected, would indicate that another oil is present, and the most likely suspect oil would be soyabean. The tocopherol profile is, however, usually only of use as confirmation, together with other analyses.

Many other analyses can be useful. As has already been stated, triglyceride analysis can determine trilinolein in oils where it should not be present, but for triglyceride and most other analyses there is little information available at present as to the natural range in the oils, and so many conclusions can only be tentative. A summary of the most likely initial indications of adulteration and their cause is given in Table 1.2.

There are some tests particular to specific oils, because of a peculiarity of that oil. Thus sesame oil is the only oil to contain sesamol and its derivatives, detected by HPLC (Raie and Salma, 1985). Unrefined groundnut oil would be expected to contain cyclopropene fatty acids, which can be detected by the Halphen test, or, more specifically and quantitatively, by HPLC or GC of the methyl esters. However, these tests are usually not very satisfactory for refined oils, as the levels of the components can be significantly lowered by refining. Indeed cyclopropene fatty acids should be completely removed from groundnut oil by refining. This is a good thing for health, as the acids are potent desaturase

Table 1.2 Indicatory signs and likely causes of adulteration

Indication[a]	Cause
1 High C16:0	Palm oil or fraction present; palm oil and palm olein is also likely to raise oleic acid, while stearin is not so likely to do so
2 Presence of significant levels of C8:0, C10:0, C12:0 and C14:0	Presence of palm kernel of coconut oil (or babassu); check *trans*-acids to see if hydrogenated; milk fat will also show these acids (and C4:0), but the amount will be steadily increasing with chain length, while the other oils have peak at C12:0
3 Presence of C22:0 and C24:0 (with little C22:1)	Groundnut oil
4 Presence of C22:1	High erucic rapeseed (not normally considered an edible oil); could also be mustard oil
5 High C18:3	Presence of soyabean or rapeseed; check for brassicasterol
6 High C18:2	Presence of sunflower, safflower, grapeseed, maize, cottonseed. Soyabean and rapeseed also, but check level of C18:3
7 High C18:1	High oleic sunflower and safflower; olive unlikely; palm olein will also raise C16:0; lard and tallow may also have high C18:0 and C14:0 is likely to be slightly elevated; check cholesterol
8 High C18:0 (low *trans* acids)	Lard or tallow
9 Brassicasterol	Rapeseed (or other *Brassica* species)
10 γ-Tocopherol	Possibly corn or soyabean present
11 Tocotrienols present	Palm or maize oil

[a]By comparison with expected range for the oil.

inhibitors and would not be good for the consumer in large amounts; however, this does mean that detection is not possible for refined oils.

1.6 Adulteration of specialist oils

There are a number of minor oils, all of high value, most of which are marketed mainly either for medical purposes or for their flavour. Olive, evening primrose, borage, fish oils and cocoa butter are described elsewhere. Others include hazelnut, walnut, macadamia, almond, apricot, pumpkin, poppy-seed and rice bran oils. The process of testing for authenticity of these oils should be approached in the same way as for the bulk oils above, i.e. fatty acid profile, sterols, tocopherols and triglyceride composition. However, there is little generally available published material on the ranges of values to be expected

Table 1.3 Sources of information on specialist oil composition

Oil	Reference
Selection of oils	Badolato *et al.*, 1987; Beuchat and Worthington, 1978; Carpenter *et al.*, 1976; Colombini *et al.*, 1979; Coors, 1991; Farines *et al.*, 1986; Fedeli, 1983; Filsoof *et al.*, 1976; Gargano, 1981; Gertz and Herman, 1982; Ghaleb *et al.*, 1991; Gombos and Woidich, 1987; Homberg and Bielefeld, 1985; Itoh *et al.*, 1974; Jeong *et al.*, 1974; Kamel and Kakuda, 1992; Lazos, 1991; Slover *et al.*, 1983; Tekin and Velioglu, 1993; Zlatanov *et al.*, 1999
Almond oil	Aitzetmuller and Ihrig, 1988; Dugo *et al.*, 1979; Gutfinger and Letan, 1973; Mehran and Filsoof, 1974; Salvo *et al.*, 1980, 1986; Saura-Calixto *et al.*, 1985
Avocado oil	Itoh *et al.*, 1975, 1976; Joseph and Neeman, 1982; Kapseu and Parmentier, 1997; Lozano *et al.*, 1993; Martinez-Nieto *et al.*, 1988; Petronici *et al.*, 1978; Poiana *et al.*, 1999; Sciancalepore and Dorbessan, 1981, 1982; Turatti and Canto, 1985
Brazil nut oil	Lima and Goncalves, 1997; da Silva *et al.*, 1997; Tateo, 1971
Cherry seed oil	Bernado-Gil *et al.*, 2001; Comes *et al.*, 1992; Farrohi and Mahran, 1975; Lotti *et al.*, 1970
Hazelnut oil	Contini *et al.*, 1991; Gargano *et al.*, 1982; Parcerisa *et al.*, 1993
Peach kernel oil	Lotti and Anelli, 1969; Rahma and Abd El Aal, 1988
Pumpkin seed oil	Markovic and Bastic, 1976; Tsuyuki *et al.*, 1985
Rice-bran oil	Norton, 1995; Rogers *et al.*, 1993

for these oils, in particular with respect to compositions varying with source. An example of this is the sterol composition of almond oil, the USA version of which apparently differs in composition from the European values. A good general source of information is an American Oil Chemists' Society publication (AOCS, 1997). Some other literature sources are listed in Table 1.3.

1.7 Oils derived from genetically modified plants

There is much controversy, at least in Europe, concerning genetic modification of plants. The three major crops affected so far are maize, soyabean and rapeseed. All of these, in addition to their other uses, are sources of oil. The reasons for modification in all these cases are related to herbicide tolerance and resistance to insects. For the varieties generally available at present, there is no known difference from non-modified strains with respect to fatty acid composition, oil yield, tocopherol level, or the level of any other minor oil constituent.

However, other varieties are in development where the composition of the oil will be deliberately affected, and in some cases these are ready for general application. Versions of soyabean aimed at increasing the levels of oleic acid, and versions of rapeseed high in lauric acid, are in development. Whether or not there is any effect on the oil composition, many consumers are requesting that oils from genetically modified sources be so labelled. Although this is not regarded in the USA as very important, in Europe regulations have come into force (EC Council, 1998) that require all foods containing protein or DNA from GM maize or soyabean to be clearly labelled. Accidental contamination of a non-GM crop with a GM variant of up to 1% is allowed.

In practical terms, at present, testing GM oils is in most cases impossible. Testing for GM crops involves looking for either foreign DNA or foreign proteins. Proteins, and to a much lesser extent DNA, are damaged by strong heat, extremes of pH and enzymic action. All refined oils, which make up the majority of world trade, are heat treated during the refining process, can be treated with acids and/or alkalis, and are then subject to filtering. Protein and DNA are likely to have been destroyed and any traces remaining will have been precipitated and filtered out. At the present state of knowledge, even using the technique of polymerase chain reaction (PCR) amplification of the DNA by a factor of several billion, there is little or no likelihood of detection of DNA in a fully refined oil (Hellebrand *et al.*, 1998). The sensitivity of PCR amplification can be further increased by various means, but increasing the sensitivity carries a concomitant risk of generating false-positive results. This is because PCR is an amplification technology and is vulnerable to cross-contamination of samples. Therefore, it is difficult to show that PCR results on refined oils are in fact a property of the oil sample and not background 'noise' in the analytical system. This problem makes PCR analysis of refined oil extremely difficult to conduct successfully.

With non-refined oils it is sometimes possible to extract intact DNA from them, apply PCR amplification, and then examine the product by gel electrophoresis. If any GM component is present, this would then be detectable. However, oils that are sold non-refined are not generally the oils that are being produced in GM forms. It is possible that in future there may GM olive, hazelnut, walnut or other non-refined oils, and it is likely that these could be tested in this way. One other oil product that can be tested is lecithin. Where there has been severe heat damage, it is not always possible to extract suitable DNA from the product, but, with samples that have been only mildly treated, it is often possible to achieve accurate results by PCR. Each case has to be evaluated separately depending on the condition of the sample, and the quality of the methods used.

When products arrive on the market that deliberately have different oil compositions, the above strictures will still apply. Unless there are traces of DNA in fully refined oils below the present level of detection, and a procedure is developed which enables DNA to be detected at these low levels, then it does not appear that it will be possible to detect highly refined GM oils by detection of 'foreign' DNA or its first generation product, i.e. protein. There remains the possibility that lipid component changes in the plant may be detected. This would not apply to many simpler changes that have been obtained. The decrease of linolenic acid and increase in oleic acid in soyabean or rapeseed can, and has been, obtained by normal breeding, in addition to those procedures using GM methods. However, versions of rapeseed containing 50% lauric acid would stand out, but these types of oils would be aimed at the non-food technical market, where there would be little interest in GM testing. Some of the minor components such as terpene alcohols, sterols or hydrocarbons could provide a method of detection. Similarly, in olive or other specialist oils there might be changes in flavour components, but oils in this class would be unrefined. As stated above, it is likely that normal DNA analysis techniques would be applicable to unrefined oils. No information is available on analysis of minor components of this type and it is likely that any differences between GM and other oils would not be of much significance.

In conclusion, it seems probable that, with the exception of unrefined oils and lecithin, there is no likelihood of detection of the presence of GM oils in most of the vegetable oils traded throughout the world in the foreseeable future. The only way to ensure that oils are authentic would be a complex, comprehensive and expensive scheme inspection and accreditation scheme, including the analysis of all batches of the source oilseed. This does not seem a likely possibility.

1.8 Organic and non-refined oils

Organic foods are gaining an increasing share of the market. The main focus of the firms marketing them was, initially, the fresh food market. Now the firms producing organic products are aiming at wider markets, including oils and fats. In order to gain organic status, a food has to satisfy various criteria. These can vary with source country, some regulatory authorities being more strict than others, but European legislation is briefly described by Shukla (2001). Organic-labelled produce avoids the use of manufactured fertilizers, pesticides and growth regulators, and should rely on crop rotation, animal and plant manure, hand weeding and biological pest control. Manufacturing processes such as irradiation, hydrogenation, fumigation and genetic modification should not be carried out on products sold as 'organic'. Foods labelled as 'organic' must be made from 95% organic ingredients, or can be labelled 'Made with organic ingredients' if the level is >70% and is listed on the label.

It can be seen that testing for an 'organic' oil could be difficult or impossible. Hydrogenation would be easy to detect but, in the present climate of opinion, is less likely to be used even in non-organic oils. It is possible to test for pesticides, but finding traces would not necessarily invalidate the organic status of the oil, as some pesticides, particularly the polychlorinated ones, persist for many years, and their presence, at low levels at least, would merely show that pesticides had been used on the land at some time, possibly years ago. In any case, intense refining could probably remove any pesticides present, at a cost. The cost would probably be justified economically to a rogue processor because of the premium on organic produce. The difficulties of ensuring an oil is not GM have already been described above. The remainder of the requirements listed above would be impossible to prove chemically. The claims of higher levels of vitamins and minerals in organic produce, whether or not true, could not be used, as they would probably still be within the generally accepted range found in the crops. Thus the only way to guarantee organic status in oils is by a guaranteed, policed, certificated, monitoring scheme. These are already in place for certain specialized oils, and, providing they are adhered to, do guarantee the validity of the product. This does come at a cost. Depending on the exact circumstances the cost could be 2–20% extra on top of the normal cost, this being in addition to the extra costs of organic production.

Non-refined oils are easier to detect and authenticate than organic oils, though the absence of a small percentage of some refined oil in the product would be difficult to prove. Where unrefined palm oil was adulterated with rapeseed has already been described above. Where an oil is authentic as to its source, but possibly at least partially refined, in order to check its authenticity, it would then be necessary to build up a database of the expected ranges of values for refined and non-refined oils. Chemical and physical techniques that should be checked are:

1. Colour. The colour of fully-refined oils is less intense than non-refined oils. Acid and alkaline treatment, and the use of bleaching earths, would all be expected to affect the colour of the oil. The method used could be spectrophotometric or colorimetric.
2. Acidity. Steam stripping, bleaching and alkali treatment would all reduce free fatty acids in the oil.
3. Presence of sterol hydrocarbons (stigmastadiene, etc.). These dienes are not present in completely unrefined oils, and are supposedly present in significant amounts in all fully refined oils, particularly those treated with strong bleaching earths or strongly deodorized (Mennie *et al.*, 1994). The test is very sensitive, and already specified in the EEC for checking purity of extra virgin olive oils. However laboratory studies have claimed that oil bleached under mild conditions does not produce enough stigmastadiene to exceed the 0.15 mg/kg limit when blended with extra virgin olive oil (Serra-Bonvehi *et al.*, 2001)
4. Tocopherols. Refining, particularly steam stripping and bleaching, does cause loss of tocopherols. However they can be added back, though whether this would be worthwhile is questionable.
5. Fatty acid composition. High temperature processes can cause formation of *trans*-acids, particularly of *trans*-octadecenoic acids. The levels are quite small, but can be detected by GC on suitable columns (BPX70, CpSil88, etc.). Of particular relevance is the fact that these, once formed,

cannot be removed by any normal refining process. These are not always formed when lower temperature processes are used and so, while their presence gives rise to a reasonable implication that refining has occurred, their absence is no guarantee that the opposite is true.

6. Triglyceride dimers in oil. Dimers, together with polymers, are formed during heating of oils. It is claimed that unrefined oils contain no detectable (<0.5%) dimers and that, if a detectable level is found, this may be caused by adulteration with a refined oil (Gertz and Klostermann, 2000). This method would probably only detect gross adulteration and not low levels.
7. Conjugated fatty acids. These can be formed by bleaching from linoleic or linolenic acid and it is claimed that detection by UV absorption, or by HPLC if levels are low, indicate bleaching of the oil (Gertz, 1991). The method would probably only be of use if a high proportion (>30%) of the oil was refined.
8. Formation of sterol dimers. These are apparently formed during bleaching and can be used to detect this process (Schulte and Weber, 1991).

In addition to the above tests, which attempt to detect refining, there are other tests that indicate solvent extraction of the oil. As many oils are solvent extracted, and because solvent extraction usually produces an oil unsuitable for culinary use without refining, then tests for solvent extracted oil are also relevant. These are:

1. Waxes
2. Aliphatic alcohols
3. Terpene alcohols

These tests are, or have been, used to check for adulteration of extra virgin olive oil (EC Council, 1991; Ntsorankoua *et al.*, 1994). A quicker method for looking for terpene alcohols has been described (Blanch *et al.*, 1998), while esters in waxes have also been examined further (Bianchi *et al.*, 1994). In both of these cases the oil examined was olive but it is likely that this would be useful, though possibly looking for different components, with other oils.

Many of the above tests except the last could be falsified by suitable additions, e.g. free fatty acid, suitable tocopherols, colour, but this would not be very likely, due to the cost relative to the limited market for these oils and the expected return.

1.9 Authentication in the future

Undoubtedly the techniques available to the oil chemist for checking the authenticity of oils will increase in number and sensitivity. It is likely that variations in trace components, including some such as polyphenolics, hydrocarbons and terpenes, which are not at present normally investigated, will enable further information to be obtained as to the source of the oil. This will probably include the area of production for at least some of the oils. Chemometric techniques will undoubtedly be of considerable use in analysing the data. It is already claimed that hydrocarbon (Webster *et al.*, 2000) and phenols and volatiles (Ranalli *et al.*, 2000) can be used to determine the area from which olive oils originate, while DNA patterns may enable different variants of olives to be detected in virgin oils (Pasqualone and Caponio, 2000) This will have to confirmed, and an agreed database set up before legislation is introduced to use these values, to authenticate the source areas of an olive oil. The reason that consumers claim to prefer certain oils to others is usually its flavour or the flavour it gives to the food it is used in. Attempts have been made to correlate perceived flavour and volatiles with area of production (Aparicio *et al.*, 1997) and in the future this will certainly be an important area of research. Taste panels are difficult to train, and there is often difficulty in producing reproducible results with different panels in different parts of the world. Possibly, when enough information has been collected, chemical analysis of volatiles will become the definitive method for assessing the quality of an olive oil, or indeed any specialist oil the choice of which depends on flavour.

A problem that has yet to be fully solved is the detection of hazelnut oil adulteration of olive oil. The fatty acid and sterol compositions are very similar and it has been suggested that this is a major problem. Fully refined hazelnut oil would be expected to contain stigmastadiene, but non-refined oils would not, though it might be expected to give an addition to the flavour volatiles. Some extra virgin oils do contain a nutty note. Whether this is due to natural olive components or originates from hazelnuts will only be certain when a robust method is available to test the oils. Several methods have been suggested (Blanch *et al.*, 1999, 2000; Breas *et al.*, 1998; Gordon *et al.*, 2001; Mannina *et al.*, 1999; Mariani *et al.*, 1999, 2001; Royer, 1999a,b; Zamora *et al.*, 2001) and the EC has recently called for proposals for funding a project addressed to finding the best method of analysis.

One aspect of authentication that will need to be addressed at some time is the question ‘What is an authentic oil?’ Even before genetically manipulated (GM) crops were developed, traditional breeding had produced oils that were different in composition to that expected a few years previously. In 1970 the proposed *Codex Alimentarius* range for erucic acid in rapeseed oil used for edible consumption was 30–60% (O’Connor and Herb, 1970). Today, after the erucic scares of the 1960s and 1970s, an edible rapeseed oil is likely to have $<1\%$ and, if an oil was presented for edible consumption with anything like the typical value for 1970, it would be rejected. Similarly, high oleic sunflower and safflower oils would not have been recognized. Now they are incorporated into the regulations as different oils. Enzyme processing can increase the yield of ‘good’ olive oil (Ranalli *et al.*, 2000, 2001). With the possibility of GM crops giving oils to the composition as required, and allowing for the fact, already discussed above, that detection of GM oils will remain very difficult, if not impossible, then perhaps it will become impossible to 'authenticate' an oil in the traditional sense, in that it may be possible to duplicate the characteristics of a specified oil from a completely different species of plant.

If that does occur, then the present system of classification of oils may be impossible to police, and a modified system may become necessary. Perhaps the sale and perceived value of oils will necessarily become dependent on the performance, not the source of the oil. With 'bulk' oils such as palm, peanut, sunflower, safflower, sesame, soya, rapeseed, corn, fish, and animal fats and oils, the fatty acid composition will obviously be important for health reasons. If the oil is to be used for frying then the frying properties will be important. In the case of palm products the physical properties and minor components such as carotenoids will be defined. Similarly animal fats will be judged mainly on physical behaviour and effect on the product in which they are used. In all cases the oxidative and stability of the oil will have to be defined. Sesame is a very stable oil, and thus its stability, together with its low level of linolenic acid, would be its major attribute, except for toasted sesame, which would probably be classed as a specialist oil. Already most baking fats sold to the public are blends developed to give the best performance, with no mention on the pack as to the source. If a bulk oil of this type had the desired chemical composition, stability and cooking behaviour, then perhaps the source would not be a matter of concern.

Specialist oils such as olive, hazelnut, walnut, almond, cocoa butter, butter, etc., could be similarly judged by desirable properties. In most cases flavour volatiles would probably be the major attribute to be tested. With others, such as cocoa butter, it would be a combination of mainly physical properties, but also some flavour properties. It should be emphasized that at present there is nowhere near enough information available on the desirable and undesirable components of these specialist oils. Olive oil, for example, contains many flavour components and polyphenolics, which have a considerable bearing on the flavour and keeping properties of the oil, while knowledge of the flavour of most of the specialist oils is still nowhere near complete. In some ways products such as this are already being developed. Mixtures of sunflower and olive oil are already being sold in the UK under proprietary names, though there is no specification as to flavour. Spreads with names containing variants of '-oliv-', are also common. Some use only refined oil and contain only 20% of this (out of 59% total fat). Due to ignorance by consumers, and lack of proper controls, many think the spreads contain extra virgin oil and no other. If it were possible to analyse the volatiles and non-volatiles of specialist oils satisfactorily then the flavour of the product could be the main specification, and not a series of tests that specifies the source, but does not necessarily specify the quality. In this case consumers would be getting a guaranteed flavour quality, though, admittedly, the oil might contain some other source ingredients. If worries were encountered that chemically manufactured flavour components were being used, then this might possibly be checked using carbon isotope ratios, as has been suggested for detecting artificially flavoured mustard oil (Remaud *et al.*, 1997). However, if a system of desirable attributes such as this were able to be introduced for flavour vegetable oils, then, for the foreseeable future, it is likely that there would little change in the product, but merely a better check on specification for the consumer. This would, of course, result in an increase in the number of grades of oil.

There will be many, particularly those involved with trade in particular oils, who will be horrified by these suggestions. They should remember that these suggestions are put forward merely as a *possible* solution to *possible* developments in oilseed breeding. If traditional methods to 'authenticate' oils break down because oil plants can be manipulated to produce a product that cannot be differentiated from another product, then the only way to ensure the exact source of an oil, and all its component parts, would be certification. This may be possible with expensive specialist oils, but is unlikely to be practicable with bulk oils. Even for specialist oils, the cost might be a problem. It would not be practicable to destroy oil which did not meet the certification. This would have to be sold at a lower price, and would be likely to interfere with sales of the genuine product. Because the flavours of oils are complex, and still partly unknown, it would not at present be possible to completely define the flavour of an oil by determining the composition of the volatiles and important non-volatiles. This is only a suggestion for the future. It would be even longer before it was economic to duplicate the composition artificially. If this were to become possible, then it would undoubtedly be done surreptitiously in any case. By defining the oil by its performance characteristics, one could ensure that the customer did not lose out. If this definition of characteristics is performed fully, then this procedure could work. The alternative would be an expensive bureaucratic certification system (which might be preferred by the EC but no-one else), or a system where one cannot be sure that any product is what it is claimed to be.

References

Aitzetmuller, K. and Ihrig, M. (1988) Fatty acid composition of almond oil, a critical discussion. *Fat Sci. Technol.*, **90**, 464–470.

Anon (1856) *The Tricks of the Trade in Adulteration of Food and Physic: With Directions for their Detection and Counteraction*, David Bogue, London, p. 74.

AOCS (American Oil Chemists' Society) (1990) *Official Methods and Recommended Practices of the American Oil Chemists' Society*, 4th edn. American Oil Chemists' Society.

AOCS (American Oil Chemists' Society) (1997) *Physical and Chemical Characteristics of Oils, Fats and Waxes*, 2nd edn. American Oil Chemists' Society.

Aparicio, R., Morales, M.T. and Alonso, V. (1997) Authentication of European virgin olive oils by their chemical compounds, sensory attributes, and consumers' attitudes. *J. Agric. Food Chem.*, **45**, 1076–1083.

Badolato, E.S.G., de Maio, F.D., Lamardo, L.C.A. and Zenebon, O. (1987) Natural oils: verification of their quality by gas liquid chromatography. *Rev. Inst. Adolfo Lutz*, **47**, 87–95.

Bastlins, L.J. (1970) Composition of lard fatty acids from pigs given beef fats in their diet. *J. Sci. Food Agric.*, **21**, 576–578.

Bernado-Gil, G., Oneto, C., Antunes, P., Rodriquez, M.F. and Empis, J.M. (2001) Extraction of lipids from cherry seed oil using supercritical carbon dioxide. *Eur. Food Res. Technol.*, **212**, 170–174.

Beuchat, L.R. and Worthington, R.E. (1978) Fatty acid composition of tree nut oils. *J. Food Technol.*, **13**, 355–358.

Bianchi, G., Tava, A., Vlahov, G. and Pozzi, N. (1994) Chemical structure long chain esters from 'sansa' olive oil. *J. Am. Oil Chem. Soc.*, **71**, 365–369.

Biedermann, M. and Grob, K. (1996) Detection of desterolized sunflower oil in olive oil through isomerized Δ^7-sterols. *Z. Lebensm. Unters. Forsch.*, **202**, 199–204.

Blanch, G.P., Villen, J. and Herraiz, M. (1998) Rapid analysis of free erythrodiol and uvaol in olive oils by coupled reversed phase liquid chromatography—gas chromatography. *J. Agric. Food Chem.*, **46**, 1027–1030.

Blanch, G.P., Caja, M.M., Ruiz del Castillo, M.L. and Herraiz, M. (1999) A contribution to the study of the enantiomeric composition of a chiral constituent in hazelnut oil used in the detrection of adulterated olive oil. *Eur. Food Res. Technol.*, **210**, 139–143.

Blanch, G.P., Caja, M.M., Leon, M. and Herraiz, M. (2000) Determination of (E)-5-methylhept-2-en-4-one in deodorised hazelnut oil. Application to the detection of adulterated olive oils. *J. Sci. Food Agric.*, **80**, 140–144.

Breas, O., Guillou, C., Reniero, F., Sada, E. and Angerosa, F. (1998) ^{17}O Measurement by continuous flow pyrolysis/isotope ratio mass spectrometry of vegetable oils. *Rapid Commun. Mass Spectrom.*, **12**, 188–192.

Carpenter, D.L., Lehman, J., Mason, B.S. and Slover, H.T. (1976) Lipid composition of selected vegetable oils. *J. Am. Oil Chem. Soc.*, **53**, 713–718.

Christie, W.W. (1986) The positional distribution of fatty acids in triglycerides. *Analysis of Oils and Fats* (eds. R.J. Hamilton and J.B. Rossell), Elsevier Applied Science, Essex, pp. 313–339.

Codex Alimentarius Commission (1997) *Proposed Draft Standard for Named Vegetable Oils*. FAO/WHO Food Standards Program.

Colombini, M., Vanoni, M.C. and Amelotti, G. (1979) Oil of walnut, hazelnut, almond, avocado: sterol composition. *Riv. Ital. Sost. Grasse*, **56**, 392–393.

Contini, M., de Santis, D. and Anelli, G. (1991) Distribution of fatty acids in glycerides and in free fatty acid fraction of hazelnut oil. *Riv. Ital. Sost. Grasse*, **68**, 405–411.

Coors, U. (1991) Utilization of tocopherol pattern for recognition of fat and oils adulterations. *Fette Seif. Anstrichm.*, **93**, 519–526.

Comes, F., Farines, M., Aumeles, A. and Soulier, J. (1992) Fatty acids and triglycerides of cherry seed oil. *J. Am. Oil Chem. Soc.*, **69**, 1224–1227.

da Silva, W.G., Cortesi, N. and Rovellini, P. (1997) The Brazilian nut. 2. Lipids: the chemical structure. *Riv. Ital. Sost. Grasse*, **74**, 311–314.

Desbordes, S., Morin, O. and Prevot, A. (1983) Study of occurrence of brassicasterol in sunflower oil and variations in sterol composition of French sunflower seeds. *Ann. Falsif. Exp. Chim. Toxicol.*, **76**, 151–169.

Dugo, G., d'Alcontres, I.S., Controneo, A. and Salvo, F. (1979) Composition of almond oil. 1. Fatty acids, hydrocarbons and sterols of three varieties of Sicilian sweet almonds. *Riv. Ital. Sost. Grasse*, **56**, 201–203.

EC Council (1991) *Characteristics of Olive Oil and Olive Residue Oil and on the Relevant Methods of Analysis*, Regulation 2568/91.

EC Council (1998) *Labelling Regulations for GM Maize and Soya*, Regulation 1139/98.

Farines, M., Soulier, J. and Comes, F. (1986) Study of the glyceride fraction of the kernel oils from some *Rosaceae*. *Rev. Franc. Corps Gras*, **33**, 115–117.

Farrohi, F. and Mahran, M. (1975) Oil characteristics of sweet and sour cherry kernels. *J. Am. Oil Chem. Soc.*, **52**, 520–521.

Fedeli, E. (1983) Miscellaneous exotic oils. *J. Am. Oil Chem. Soc.*, **60**, 404–406.

Filsoof, M., Mehran, M. and Farrohi, F. (1976) Determination and comparison of oil in Iranian almond, apricot and peach nuts. *Fette Seif. Anstrichm.*, **78**, 150–151.

Firestone, D. (2001) Assuring the integrity of olive oil products. *J. Assoc. Off. Anal. Chem.*, **84**, 176–180.

Flor, R.V., le Tiet, H. and Martin, B.D. (1993) Development of high-performance liquid chromatography criteria for determination of grades of commercial olive oils. *J. Am. Oil Chem. Soc.*, **70**, 199–203.

FOSFA (1994) *Guideline Specifications*, 2nd edn. Federation of Oil Seed and Fat Associations International Ltd., London.

Gargano, A. (1981) Fatty acid and sterol composition of oils used in foods. Ind. Aliment., **20**, 510–513.

Gargano, A., Magro, A. and Pellegrino, M. (1982) Caratterische chimiche dei frutti di alcune delle principale cultivar di noccoile. *Ind. Aliment.*, **21**, 15–16.

Gertz, C. (1991) Native and unrefined edible oils and fats. *Fett Wissen. Technol.*, **93**, 545–548.

Gertz, C. and Hermann, K. (1982) Analysis of tocopherols and tocotrienols in foods. *Z. Lebensm. Unters. Forsch.*, **174**, 390–394.

Gertz, C. and Klostermann, S. (2000) A new analytical procedure to differentiate virgin and non-refined from refined vegetable oils. *Eur. J. Lipid Sci. Technol.*, **102**, 329–336.

Ghaleb, M.L., Farines, M. and Soulier, J. (1991) Composition chimique deshuiles de graines de citrouille, courge, melon. *Rev. Franc. Corps Gras*, **38**, 17–22.

Gombos, J. and Woidich, H. (1987) Influence of production and processing on the major and minor constituents of vegetable oils. *Ehnahrung*, **11**, 539–545.

Gordon, M.H., Covell, C. and Kirsch, N. (2001) Detection of pressed hazelnut oil in admixtures with virgin olive oil by analysis of polar components. *J. Am. Oil Chem. Soc.*, **78**, 621–624.

Grob, K., Biedermann, M. and Bronz, M. (1994) Results of a control of edible oils: frauds by admixtures, contaminations. *Mittel. Geb. Lebens. Hyg.*, **85**, 351–365.

Gutfinger, T. and Letan, A. (1973) Detection of adulteration of almond oil with apricot oil through determination of tocopherols. *J. Agric. Food Chem.*, **21**, 1120–1123.

Gutfinger, T. and Letan, A. (1974) Studies of unsaponifiables in several vegetable oils. *Lipids*, **9**, 658–663.

Hassell, A.H. (1861) *Adulterations Detected or Plain Instructions for the Discovery of Frauds in Food and Medicine*. Longman, Green, Longman and Roberts, London.

Hellebrand, M., Nagy, M. and Moersel, J.T. (1998) Determination of DNA traces in rapeseed oil. *Z. Lebensm. Unters. Forsch. A*, **206**, 237–242.

Hilditch, T.P. and Williams, P.N. (1964) *The Chemical Constitution of Natural Fats*, 4th edn. Chapman & Hall, London.

Homberg, E. and Bielefeld, B. (1985) Free and bound sterols in vegetable fats. *Fette Seif. Anstrichm.*, **87**, 61–64.

Itoh, T., Tamura, T. and Matsumoto, T. (1973) Sterol composition of 19 vegetable oils. *J. Am. Oil Chem. Soc.*, **50**, 122–225.

Itoh, T., Tamura, T. and Matsumoto, T. (1974) Sterols and methylsterols in some tropical and subtropical vegetable oils. *Oleagineux*, **29**, 253–258.

Itoh, T., Tamura, T., Matsumoto, T. and Dupaigne, P. (1975) Studies on avocado oil, especially the sterol fraction of the unsaponifiable. *Fruits*, **30**, 687–695.

Itoh, T., Tamura, T., Matsumoto, T. and Dupaigne, P. (1976) Studies on avocado oil, especially the sterol fraction. 2. The 4-monomethyl sterol fraction. *Fruits*, **31**, 473–481.

Iverson, J.L. (1972) Gas-liquid chromatographic detection of palm kernel and coconut oils in cacao butter. *J. Assoc. Off. Anal. Chem.*, **55**, 1319–1322.

Iverson, J.L., Eisner, J. and Firestone, D. (1965) Detection of trace fatty acids in fats and oils by urea fractionation and gas-liquid chromatography. *J. Assoc. Off. Anal. Chem.*, **42**, 1063–1068.

James, A.T. and Martin, A.J.P. (1952) Gas-liquid partition chromatography: the separation and microestimation of volatile fatty acids from formic acid to dodecanoic acid. *Biochem. J.*, **50**, 679–690.

Jeong, T.M., Itoh, T., Tamura, T. and Matsumoto, T. (1974) Analysis of sterol fractions from twenty vegetable oils. *Lipids*, **9**, 921–927.

Joseph, D. and Neeman, I. (1982) Characterisation of avocado oil by polyalcoholic compounds in the unsaponifiable matter. *Riv. Ital. Sost. Grasse*, **59**, 279–284.

Kamel, B.S. and Kakuda, Y. (1992) Characterisation of the seed oil and meal from apricot, cherry, nectarine, peach and plum. *J. Am. Oil Chem. Soc.*, **69**, 492–494.

Kapseu, C. and Parmentier, M. (1997) Fatty acid composition of some vegetable oils from Cameroon. *Sci. Aliment.*, **17**, 325–331.

Kathmandu Post (2001) *Food Adulteration*, Editorial 7 May 2001. Mercantile Communications, Kathmandu.

Kelly, S., Parker, I., Sharman, M., Dennis, J. and Goodall, I. (1997) Assessing the authenticity of single seed oils using fatty acid stable carbon isotope ratios. *Food Chem.*, **59**, 181–186.

Kurtzweil, P. (1999) *Fake food fight*, FDA Consumer, March-April US Food and Drug Administration. http://vm.cfsan.fda.gov/~dms/fdfake.html

Lanuzza, F. and Micali, G. (1997) On-line LC-GC-FID determination of Δ^7-stigmasatenol and Δ^8-1,4-stigmasatenol in edible oils. *Riv. Ital. Sost. Grasse*, **74**, 509–512.

Lazos, E.S. (1991) Composition and oil characteristics of apricot, peach, and cherry kernels. *Grasas Aceites*, **42**, 127–131.

Lewkowitsch, J. (1895) *Chemical Technology and Analysis of Oils, Fats and Waxes*, 1st edn, Macmillan, London.

Lewkowitsch, J. (1904) *Chemical Technology and Analysis of Oils, Fats and Waxes*. 3rd edn, Macmillan, London.

Lima, J.R. and Goncalves, L.A.G. (1997) Tocopherol quantification in oils (corn, soy bean, Brazil nut, and cashew nut) by high performance liquid chromatography. *Alimenos e Nutricao*, **8**, 65–73.

Lotti, G. and Anelli, G. (1969) The seed oils of *Prunus persica* Sieb. *Riv. Ital. Sost. Grasse*, **46**, 110–114.

Lotti, G., Pisano, G., Anelli, G. and Baragli, S. (1970) Seed oils in some varieties of cherry. *Riv. Sci. Aliment.*, **16**, 248–253.

Lozano, Y.F., Mayer, C.D., Bannon, C. and Gaydou, E.M. (1993) Unsaponifiable matter, total sterol and tocopherol contents of avocado oil varieties. *J. Am. Oil Chem. Soc.*, **70**, 561–565.

Mannina, L., Patumi, M., Fiordiponti, P., Emanuele, M.C. and Segre, A.L. (1999) Olive and hazelnut oils: a study by high field ^{1}H NMR and gas chromatography. *Ital. J. Food Sci.*, **11**, 139–149.

Mariani, C. and Venturini, S. (1997) Characterization of desterolized high oleic acid sunflower oil in olive oil. *Riv. Ital. Sost. Grasse*, **74**, 489–500.

Mariani, C., Bellan, G., Morchio, G. and Pellegrino, A. (1999) Free and esterified minor components of olive and hazelnut oils: their potential utilisation in checking oil blend *Riv. Ital. Sost. Grasse*, **76**, 297–305.

Mariani, C., Bellan, G., Morchio, G. and Pellegrino, A. (2001) Some unusual minor components in hazelnut oil and olive oil. *Riv. Ital. Sost. Grasse*, **78**, 3–16.

Markovic, V.V. and Bastic, L.V. (1976) Characteristics of pumpkin seed oil. *J. Am. Oil Chem. Soc.*, **53**, 42–44.

Martel, J. (1977) Fractional crystallisation of the triglycerides of virgin olive oil. Presence of semi-drying oils. *Grasas Aceites*, **28**, 189–198.

Martinez-Nieto, L., Camacho-Rubio, F., Rodriguez-Vives, S. and Moreno-Romero, M.V. (1988) Extraction and characterisation of avocado oil. *Grasas Aceites*, **39**, 272–277.

Mehran, M. and Filsoof, M. (1974) Characterisation of Iranian almond nuts and oils. *J. Am. Oil Chem. Soc.*, **51**, 433–434.

Mennie, D., Moffat, C.F. and McGill, A.S. (1994) Identification of sterene compounds produced during the processing of oils. *J. High Resol. Chromatogr.*, **17**, 831–838.

Mitchell, J. (1848) *Treatise on the Falsification of Food*. J.B. Baillière, London.

Noah, M.M. (1844) Lard oil and kosher certification of olive oil. *The Occident and Jewish Advocate*, **11**(7), 5605.

Norton, R.A. (1995) Quantitation of steryl ferulate and p-coumaric esters from corn and rice. *Lipids*, **30**, 269–274.

Ntsorankoua, H., Artaud, J. and Gueree, M. (1994) Triterpene alcohols in virgin olive oil and refined olive pomace oil. *Ann. Falsif. Exp. Chim Toxicol.*, **87**, 91–107.

O'Connor, R.T. and Herb, S.F. (1970) Specification of fatty acid composition for identification of oils and fats. *J. Assoc. Off. Anal. Chem.*, **47**, 195A.

Parcerisa, J., Boatella, R., Codony, R., Farran, A., Garcia, J., Lopez, A., Rafecas, M. and Romero, A. (1993) Influence of variety and geographical origin on the lipid fraction of hazelnut from Spain.1. Fatty acid composition. *Food Chem.*, **48**, 411–414.

Pasqualone, A. and Caponio, F. (2000) Molecular markers in olive tree; current state and potential applications to olive oils. *Ind. Aliment.*, **39**, 1397–1402.

Petronici, C., Bazan, E., Panno, M. and Averna, V. (1978) Composition of sicilian avocado oil. *Riv. Ital. Sost. Grasse*, **55**, 260–262.

Poiana, M., Giuffre, A.M., Mincione, B., and Giuffre, F. (1999) Avocado oil. Lipidic components evolution during the ripening of the fruits of some cultivars grown in the south of Italy. *Riv. Ital. Sost. Grasse*, **76**, 257–275.

Posada de la Paz, M., Philen, R.M., Abaitua Borda, I., Sicilia Socias, J.M., Gomez de la Camara, A. and Kilbourne, E.M. (1996) Toxic oil syndrome: Traceback of the toxic oil and evidence for a point source epidemic. *Food Chem. Toxicol.*, **34**, 251–257.

Postgate, J. (1885) *Second Report from the Select Committee on Adulteration Food, Drink and Drugs: Minutes of Evidence*. British Parliamentary Papers, 3098–3134.

Rahma, E.H. and Abd El Aal, M.H. (1988) Chemical characterisation of peach nut oil an protein: functional properties, *in vitro* digestibility and amino acids profile of flour. *Food Chem.*, **28**, 31–43.

Raie, M.Y. and Salma, S. (1985) *Sesamum indicum* and *Papaver somniferum* oils. *Fette Seif. Anstrichm.*, **87**, 246–247.

Ranalli, A., Costantini, N., de Mattia, G. and Ferrante, M.L. (2000) Evaluating two kinds of centrifuged virgin oils arising from continuous olive processing. *J. Sci. Food. Agric.*, **80**, 673–683.

Ranalli, A., Malfatti, A. and Cabras, P. (2001) Composition and quality of pressed virgin olive oils extracted with a new enzyme processing aid. *J. Food Sci.*, **66**, 592–603.

Remaud, G.S. Martin, Y.L., Martin, G.G., Naulet, N. and Martin, G.J. (1997) Authentication of mustard oils by combined stable isotope analysis (SNIF-NMR and IRMS). *J. Agric. Food Chem.*, **45**, 1844–1848.

Rogers, E.J., Rice, S.M., Nicolosi, R.J., Carpenter, D.R., McClelland, C.A. and Romanczyk, L.J. (1993) Identification and quantitation of γ-oryzanol components and simultaneous assessment of tocols in rice bran oil. *J. Am. Oil Chem. Soc.*, **70**, 301–307.

Rossell, J.B. (1985) Composition of oil. *J. Am. Oil Chem. Soc.*, **62**, 221–230.

Rossell, J.B. (1991) Purity criteria in edible oils and fats. *Fat Sci. Technol.*, **93**, 526–531.

Rossell, J.B. (1994) Stable isotope ratios in establishing maize oil purity. *Fat Sci. Technol.*, **96**, 304–208.

Royer, A., Gerard, C., Naulet, N., Lees, M. and Martin, G.J. (1999a) Stable isotope characterisation of olive oils. 1. Compositional and carbon-13 profiles of fatty acids. *J. Am. Oil Chem. Soc.*, **76**, 357–363.

Royer, A., Naulet, N., Mabon, F., Lees, M. and Martin, G.J. (1999b) Stable isotope characterisation of olive oils. 2. Deuterium distribution in fatty acids studied by nuclear magnetic resonance. *J. Am. Oil Chem. Soc.*, **76**, 365.

Salvo, F., Dugo, G., Stago d'Alcontres, I., Corroneo, A. and Dugo, G. (1980) Composition of almond oil. 2. Distinction of sweet almond oil from blends with peach and apricot seed oil. R*iv. Ital. Sost. Grasse*, **57**, 24–26.

Salvo, F., Alfa, M. and Dugo, G. (1986) Almond oil composition: variation of several chemical and physio-chemical parameters during storage. *Riv. Ital. Sost. Grasse*, **63**, 37–40.

Saura-Calixto, F., Canellas, J. and Gracia-Raso, A. (1985) Characteristics and fatty acid composition of almond tegument oil—comparison with almond kernel oil. *Fette Seif. Anstrichm.*, **87**, 4–6.

Schulte, E. and Weber, N. (1991) Disteryl ethers in edible oils and fats—comparison with the conjugated trienes. *Fett Wissen. Technol.*, **93**, 517–518.

Sciancalepore, V. and Dorbessan, W. (1981) Effect of variety on fatty acid composition and glyceride structure of avocado oil. *Riv. Agri. Subtrop. Trop.*, **75**, 109–115.

Sciancalepore, V. and Dorbessan, W. (1982) Sterol composition of avocado oil. *Grasas Aceites*, **33**, 273–275.

Serra-Bonvehi, J., Soliva-Torrento, M. and Ventura-Coli, F. (2001) A laboratory study of the bleaching process in stigmasta-3,5-diene concentration in olive oils. *J. Am. Oil Chem. Soc.*, **78**, 305–310.

Shukla, V.K.S. (2001) Organic Foods: Present and Future Developments. *Inform*, **12**, 495–499.

Sloane, T.O'C. (1907) Adulteration of food. *The Catholic Encyclopaedia*, Vol. 1. Robert Appleton Co., New York (Online edition, 1999, http://www.newadvent.org/cathen/01162b.htm).

Slover, H.T., Thompson, J.R. and Merola, G.V. (1983) Determination of tocopherols and sterols by gas chromatography. *J. Am. Oil Chem. Soc.*, **60**, 1524–1528.

Spangenberg, J.E. and Ogrinc, N. (2001) Authentication of vegetable oils by bulk and molecular carbon isotope analysis with emphasis on olive oil and pumpkin seed oil. *J. Agric. Food Chem.*, **49**, 1534–1540.

Synouri-Vrettakou, S., Komaitis, M.E. and Voudouris, E.C. (1984) Triglyceride composition of olive oil, cottonseed oil and their mixtures by low temperature crystallisation and gas liquid chromatography. *J. Am. Oil Chem. Soc.*, **61**, 1051–1056.

Tan, B.K., Siew, W.L., Oh, F.C.H. and Berger, K.G. (1983) Detection of palm stearine in palm oil, in *Palm Oil Product Technology in the Eighties*. Palm Oil Research Institute of Malaysia, Kuala Lumpa, pp. 165–181.

Tateo, F. (1971) Acid composition of fat material extracted from seeds of *Bertholletia excelsa*. *Ind. Aliment.*, **10**, 68–70.

Tekin, A. and Velioglu, S. (1993) Research on some compositional properties of melon seed and bitter almond. *Gida*, **18**, 365–367.

Tsuyuki, H., Itoh, S. and Yamagata, K. (1985) Lipid and triacylglycerol compositions of total lipids in pumpkin seeds. *Nippon Shokuhin Kogyo Gakkaishi*, **32**, 7–15.

Turatti, J.M. and Canto, W.L. (1985) Unsaponifiables of avocado oil. *Bull. Inst. Technol. Aliment., Brazil*, **22**, 311–329.

von Peters, H. and Wieske, Th. (1966) Detection of traces of polybranched fatty acids in fats. *Fette Seif. Anstrichm.*, **68**, 947–950.

Webster, L., Simpson, P., Shanks, A.M. and Moffat, C.F. (2000) The authentication of olive oil on the basis of hydrocarbon concentration and composition. *Analyst*, **125**, 97–104.

Working Party on Food Authenticity (1996) *Authenticity of Single Seed Oils*. Ministry of Agriculture, Fisheries and Foods, London.

Wurzinger, J. and Hensel, G. (1969) Determination of the nature of meat from the fat-components and substances accompanying fats. *Fette Seif. Anstrichm.*, **71**, 144–151.

Yi, L.L., Kwo, L.K. and Hui, Y.H. (1993) Preliminary studies on the adulteration detection of commercial sesame oils. *J. Chinese Agric. Chem. Soc.*, **31**, 697–701.

Youk, M.C., Hsiu, J.L., Chung, W.C. and Mei, L.W. (1999) A rapid gas chromatographic method for direct determination of free sterols in animal and vegetable fats and oils. *J. Food Drug Anal.*, **7**, 279–289.

Zamora, R., Alba, V. and Hidalgo, F.J. (2001) Use of high resolution ^{13}C NMR spectroscopy for the screening of virgin olive oils. *J. Am. Oil Chem. Soc.*, **78**, 89–94.

Zlatanov, M., Ivanov, S. and Aitzetmueller, K. (1999) Phospholipid and fatty acid composition of Bulgarian nut oils. *Eur. J. Lipid Sci.Technol.*, **101**, 437–439.

2 Authentication of olive oil

Giorgio Bianchi

2.1 Introduction

This chapter presents an overview of the methods of extraction, together with descriptions and designations, of olive oils (European Community Council regulations EC 136/66 and EC 1638/98), and explains and comments on the purpose and scope of the relevant official methods of olive oil analysis according to the European Community (EC) regulation EC 2568/91 and subsequent additions and amendments (EC Council, 1991).[a]

From the points of view of both commercial value and health, the authenticity of the various categories of olive oil is of great importance. Olive oil has gained in popularity in many countries and there is a growing demand, particularly in countries where the standard of living is rising. The increased popularity of olive oil is due to its peculiar organoleptic characteristics and to its health benefits—the latter being no longer just 'alleged', but proven. Olive oils command a higher price than other vegetable oils, due to their popularity and high costs of production, and there is, consequently, a great temptation for the purveyors of fraud to produce and market sophisticated olive oils.

2.2 From olives to olive oil

Olive oil is a fatty fruit juice and, as such, is directly consumable after the appropriate, diversified, processing of olives. The Mediterranean countries, where the olive tree has been cultivated for thousands of years, are still responsible for over 90% of the world's olive oil, although olive cultivation has also been successful in Argentina, Australia, California and South Africa. The olive, a delicate pulpy fruit, is harvested by gently shaking the tree and branches with appropriate mechanical shakers and collecting the fruit in nets placed on the ground under the tree; undamaged fruit gives the highest quality oil. In the Mediterranean basin harvesting is commonly carried out just before the olives ripen, usually from October to December. The pressing of the olives can be carried out during or immediately after the harvest. Maturation of olives can be followed by observing the colour change of fruit from green to changing colour

[a]The methods and standards are revised and updated frequently. The revision is often the result of collaborative work with the International Olive Oil Council (IOOC), the *Codex Alimentarius* Joint FAO/WHO and with other national and international authorities.

to black. However, there are olive cultivars whose fruits hardly become wholly black even at complete maturation. Different cultivars, in general, have different periods of ripeness. This fact facilitates the labour of the olive grower who will be able to harvest and process large amount of olives, during an appropriate length of time according to plant processing capacity. The major reason for harvesting at different periods of ripeness is that the oil obtained from some cultivars possesses markedly different characteristics both in composition and, consequently, in organoleptic properties. For the best oil the harvested olives must be processed within two or three days, and it is common practice to mill the olives whole without removing the stones. However, recently, some producers have taken the unique approach of pitting the olives before expressing the oil, with the resulting oil differing only slightly in composition from that extracted from whole olives, although the organoleptic characteristics are quite different.

2.2.1 Extraction methods

Virgin olive oil is most commonly extracted using pressure and centrifugation systems, a less common method being the percolation system. Crushing is designed to burst the olive flesh and kernel, the droplets of oil thus running from the cells. Two types of machines are generally used for crushing: traditional stone mills and modern mechanical crushers. Beating the paste favours the separation of the liquid phases from the solid, and breaks up the oil–water emulsions. Pressing is used by small producers to separate the oil from the paste. The paste is spread over 'pressing bags' piled one above the other on press plates; the bags act as drainage tubes, allowing the oil to flow onto the lower plate of the press but holding back the solids. The obtained oil is cloudy, having a thick appearance because of the dispersed water and skin and pulp particles. Such undecanted oil is much appreciated by gourmet consumers who value its rustic, fresh aroma. Decanted clear oil is obtained by allowing the particles to settle on the bottom of the container over a 2–3 month period; as an alternative the undecanted oil can be cleaned by filtration or centrifugation. In all of these, the pressing method guarantees a high quality oil because of the short beating time and the low temperature of the entire process. Centrifugation is a method widely used in industrial plants to separate oil. Regrettably, its advantage of a greater processing capacity, compared with pressing, is tempered by the loss of important minor oil components when water is added to dilute the paste.

Since oil adheres to metal surfaces to a far greater degree than does water, it is this physical property of adherence that facilitates the recovery of the oil dispersed in the olive paste by percolation. Equipment constructed on the basis of this property is capable of extracting over 70% of the oil present in olive paste. The method is less productive than pressing and centrifuging, but the reward is a top quality oil that many producers rightly consider to be 'prime' (Bianchi, 1999; Di Giovacchino, 2000).

2.2.2 *Exhaustive extraction of olive oil: olive-residue oil*

Both the press and centrifuge solid residues, called the pomace, can be extracted still further by solvent extraction. The oil thus obtained can, after refining, be included among edible-grade olive oils.

2.3 Olive oil composition: major compound classes

The main components of almost 97–98% of whole olive oil are substances of a glyceride nature concentrated in the pulp and seed. The remaining non-acylglycerol lipid fraction is a mixture of compound classes, including alkanes, squalene, wax esters, aliphatic alcohols and aldehydes, tetracyclic (sterols) and pentacyclic triterpenes (acids, alcohols and esters), free fatty acids, vitamins, phospholipids, polyphenols and glycosides, distributed in the various parts of the fruit (Figure 2.1). The free fatty acid proportions are in the narrow 1–3% range for olive oils obtained by simply pressing the fruit, whereas these can be as high as 10–15% for solvent-extracted oils (olive-residue or 'sansa' olive oil). It is the minor compound classes, their concentrations and relative percentages, that are the determinants of olive oil characterization and commercial grading

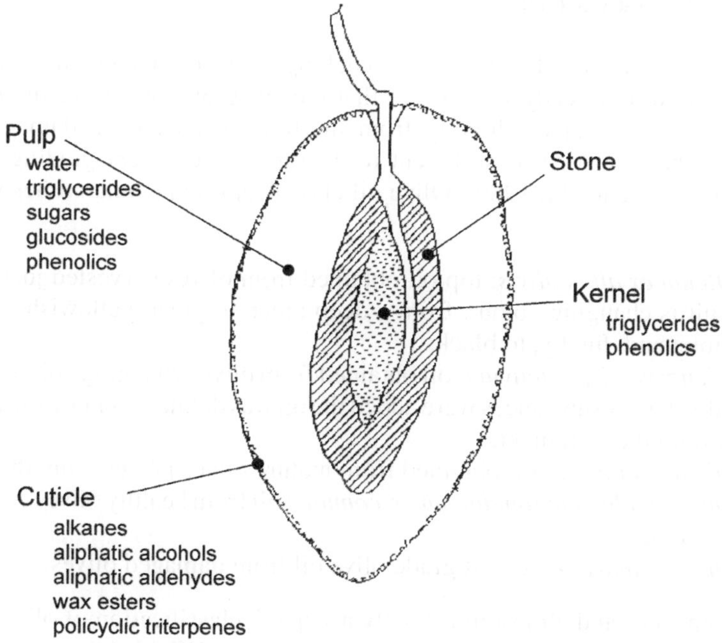

Figure 2.1 Section through a mature olive fruit showing distribution of major classes of compounds.

Table 2.1 Significance of the olive oil quality parameters

Parameters	Significance
Free fatty acids (%)	Deterioration of oils
Peroxide value (meq O_2/kg)	Presence of hydroperoxides
Halogenated solvents	Detection of harmful contamination
Phenols (mg/kg)	Antioxidants
Induction time (h)	Stability, resistance to oxidation
Chlorophyll pigments (mg/kg)	Influence on oil acceptability by consumer
K_{232}	Double bond conjugation: (i) drastic thermal and chemical
K_{270}	treatment of oil; (ii) oxidation
Panel test	Organoleptic analysis

(Bianchi and Vlahov, 1994). Furthermore, as has been pointed out in many research papers, some of these minor components are determinant factors of oil stability, as well as being relevant from the hedonistic and salutary points of view (Table 2.1). The natural concentration of these minor components in an oil can vary greatly, being related mainly to the cultivar, the stage of maturity of the fruits, the soil, the climate and also to the extraction technique adopted (Kiritsakis and Christie, 2000).

2.4 Olive oil categories

Olive oil classification has a history reaching back to ancient times. In fact the Romans were already well aware that the stage of maturity of the olives, whether they were taken directly from the tree, or picked up damaged (or not) from the ground, was sufficient to determine a different grading of the oil obtained (Cucurachi, 1989). Olive oil classification in Roman times was as follows:

1. *Oleum ex albis ulivis*: top oil obtained from olives harvested just as the fruit is changing colour, from a deep green to green-yellowish, to dark violet and, finally, to black.
2. *Oleum viride, strictivum*: oil obtained from olives changing colour; when odoriferous substances were added the oil was defined *oleum ad unguenta* and used as ointment.
3. *Oleum maturum*: oil obtained from mature olives picked from the tree.
4. *Oleum caducum*: *romanicum* or *comune*; oil from healthy olives collected from the ground.
5. *Oleum cibarium*: lowest grade olive oil from damaged olives.

The present-day, and almost universally accepted, classification of olive oils is that defined in the European Community regulation EC 136/66, amended by regulations EC 1638/98 and EC 1513/01, which set down denominations and definitions of olive oils and olive-residue oils. Marketing is in accordance with the following designations and definitions:

> 'Virgin olive oils are oils obtained from the fruit of the olive tree by solely mechanical, or other physical means under conditions, particularly thermal, that do not lead to alterations in the oil, such oil not having undergone any treatment other than washing, decantation, centrifugation and filtration. Thus there is the total exclusion of oils obtained using solvents or re-esterification processes, and any mixtures with oils of other kinds.

1. *Extra virgin olive oil* is a virgin olive oil that has a free acidity, expressed as oleic acid, of not more that 1 g per 100 g, and any other characteristics correspond to those fixed for this category in Annex I to EC regulation 2568/91 and amendments. The oil is fit for consumption "as is".
2. *Virgin olive oil* with the qualifier *fine*, a term that can be used at the production and wholesale stages, is a virgin olive oil that has a free acidity, expressed as oleic acid, of not more than 2 g per 100 g and characteristics corresponding to those fixed for this category in Annex I to EC regulation 2568/91 and amendments. The oil is fit for consumption "as is".
3. *Ordinary virgin olive oil* is a virgin olive oil that has a free acidity, expressed as oleic acid, of not more than 3.3 g per 100 g and characteristics corresponding to those fixed for this category in Annex I to EC regulation 2568/91 and amendments.
4. *Virgin lampante olive oil*, is virgin olive oil that has a free acidity, expressed as oleic acid, of more than 3.3 g per 100 g and/or characteristics corresponding to those fixed for this category in Annex I to EC regulation 2568/91 and amendments.
5. *Refined olive oil* is an olive oil obtained from virgin olive oils using refining methods; this oil has a free acidity, expressed as oleic acid of not more than 0.5 g per 100 g and characteristics corresponding to those fixed for this category in Annex I to EC regulation 2568/91 and amendments.
6. *Olive oil* is an oil that is a blend of refined olive oil and virgin olive oils, except for virgin lampante olive oil; it has a free acidity, expressed as oleic acid, of not more than 1.5 g per 100 g and characteristics corresponding to those fixed for this category in Annex I to EC regulation 2568/91 and amendments. The oil is fit for consumption "as is".
7. *Crude olive-residue oil* is the oil obtained by treating olive pomace with solvents; any oils obtained by re-esterification processes or through mixtures with oils of other kinds are totally excluded. The characteristics of the oil correspond to those fixed for this category in Annex I to EC regulation 2568/91 and amendments.
8. *Refined olive-residue oil* is the oil obtained from crude olive-residue oil by refining methods; it has a free acidity expressed as oleic acid of not more 0.5 g per 100 g and the characteristics correspond to those fixed in Annex I to EC regulation 2568/91 and amendments.

Table 2.2a Annex I—Characteristics of olive oil

Type	Acidity (%)	Peroxide value (mEq O_2/kg)	Halogenated solvents (mg/kg)	Waxes (mg/kg)	Saturated fatty acids in position 2 triglyceride (%)	Stigmasta-dienes (mg/kg)	Difference ECN42 HPLC −ECN42 calculated	K_{232}	K_{270}	K_{270} after passage over alumina	ΔK	Panel test
1. Extra virgin olive oil	≤1.0	≤20	≤0.20	≤250	≤1.3	≤0.15	≤0.2	≤2.50	≤0.20	≤0.10	≤0.01	≥6.5
2. Virgin olive oil	≤2.0	≤20	≤0.20	≤250	≤1.3	≤0.15	≤0.2	≤2.60	≤0.25	≤0.10	≤0.01	≥5.5
3. Ordinary virgin olive oil	≤3.3	≤20	≤0.20	≤250	≤1.3	≤0.15	≤0.2	≤2.60	≤0.25	≤0.10	≤0.01	≥3.5
4. Virgin lampante olive oil	>3.3	>20	>0.20	≤350	≤1.3	≤0.50	≤0.3	≤3.70	>0.25	≤0.11	—	<3.5
5. Refined olive oil	≤0.5	≤5	≤0.20	≤350	≤1.5	—	≤0.3	≤3.40	≤1.20	—	≤0.16	—
6. Olive oil	≤1.5	≤15	≤0.20	≤350	≤1.5	—	≤0.5	≤3.30	≤1.00	—	≤0.13	—
7. Crude olive-residue oil	>0.5	—	—	—	≤1.8	—	≤0.6	—	—	—	—	—
8. Refined olive-residue oil	≤0.5	≤5	≤0.20	—	≤2.0	—	≤0.5	≤5.50	≤2.50	—	≤0.25	—
9. Olive-residue oil	≤1.5	≤15	≤0.20	>350	≤2.0	—	≤0.5	≤5.30	≤2.00	—	≤0.20	—

Table 2.2b Annex I—Characteristics of olive oil

Type	Acid composition															
	Myristic (%)	Linolenic (%)	Arachiolic (%)	Eicosenoic (%)	Behenic (%)	Lignoceric (%)	Sum of the *trans*-oleic isomers	Sum of the *trans*-linoleic and *trans*-linolenic isomers (%)	Chole-sterol (%)	Brassica-sterol (%)	Campe-sterol (%)	Stigma-sterol (%)	β-sitosterol *apparent* (%)	δ-7-Stigma-stenol (%)	Total sterols (mg/kg)	Erytro-diol + uvaol (%)
1. Extra virgin olive oil	≤0.05	≤0.9	≤0.6	≤0.4	≤0.2	≤0.2	≤0.05	≤0.05	≤0.5	≤0.1	≤4.0	<Camp.	≥93.0	≤0.5	≥1000	≤4.5
2. Virgin olive oil	≤0.05	≤0.9	≤0.6	≤0.4	≤0.2	≤0.2	≤0.05	≤0.05	≤0.5	≤0.1	≤4.0	<Camp.	≥93.0	≤0.5	≥1000	≤4.5
3. Ordinary virgin olive oil	≤0.05	≤0.9	≤0.6	≤0.4	≤0.2	≤0.2	≤0.05	≤0.05	≤0.5	≤0.1	≤4.0	<Camp.	≥93.0	≤0.5	≥1000	≤4.5
4. Virgin lampante olive oil	≤0.05	≤0.9	≤0.6	≤0.4	≤0.2	≤0.2	≤0.10	≤0.10	≤0.5	≤0.1	≤4.0	—	≥93.0	≤0.5	≥1000	≤4.5
5. Refined olive oil	≤0.05	≤0.9	≤0.6	≤0.4	≤0.2	≤0.2	≤0.20	≤0.30	≤0.5	≤0.1	≤4.0	<Camp.	≥93.0	≤0.5	≥1000	≤4.5
6. Olive oil	≤0.05	≤0.9	≤0.6	≤0.4	≤0.2	≤0.2	≤0.20	≤0.30	≤0.5	≤0.1	≤4.0	<Camp.	≥93.0	≤0.5	≥1000	≤4.5
7. Crude olive-residue oil	≤0.05	≤0.9	≤0.6	≤0.4	≤0.3	≤0.2	≤0.20	≤0.10	≤0.5	≤0.2	≤4.0	—	≥93.0	≤0.5	≥2500	≥12
8. Refined olive-residue oil	≤0.05	≤0.9	≤0.6	≤0.4	≤0.3	≤0.2	≤0.4	≤0.35	≤0.5	≤0.2	≤4.0	<Camp.	≥93.0	≤0.5	≥1800	≥12
9. Olive residue oil	≤0.05	≤0.9	≤0.6	≤0.4	≤0.3	≤0.2	≤0.4	≤0.35	≤0.5	≤0.2	≤4.0	<Camp.	≥93.0	≤0.5	≥1600	>4.5

Camp., campesterol.

9. *Olive-residue oil* is an oil consisting of a blend of refined olive-pomace oil and virgin olive oils, except for virgin lampante oil. The oil has a free acidity, expressed as oleic acid, of not more than 1.5 g per 100 g and the characteristics correspond to those fixed for this category in Annex I to EC regulation 2568/91 and amendments. The oil is fit for consumption "as is" ' (Table 2.2).

2.5 Contextual meaning of words used

The definitions given in Webster's dictionary for 'authentic' and 'genuine' are: *authentic* implies being fully trustworthy as according with fact or actuality; *genuine* implies accordance with an original or a type without counterfeiting, admixture, or adulteration.

The subtlety of word-meaning proposed above needs to be confronted with the complex reality of olive oil denomination nomenclature, as detailed earlier in Section 2.4.

From the commercial or trade point of view there is no need to give any further attribute to the official oil category; for example the term 'extra virgin olive oil' is complete in itself in defining a precise type of olive oil. However, in everyday parlance, and to express a sort of general informal judgement of product quality, the adjective 'genuine' should apply only to the first three olive oils, i.e. extra virgin, virgin and ordinary virgin. On the contrary, and in agreement with what is specified in the dictionary, the word 'authentic' infers neither quality judgement nor bias; consequently when, for example, the term 'authentic olive-pomace oil' is used, the intention is to define simply and precisely only one of the nine categories of olive oil.

Misbranding. Olive oil is unique in that it is the oil from one fruit marketed in nine different categories. Seed oils, instead, are almost all oils refined from a single seed species or a mixture of different seeds. Thus, on considering fraud and adulteration practices, it is advisable in the olive oil field to avoid misunderstanding by using the following words in accordance with their clearly strict definition:

1. *alteration*: compositional molecular changes occurring in the oil and affecting the quality of the commodity (increase in free acids, formation of peroxides and derivatives, rancidity, altered ultraviolet absorption etc.)
2. *sophistication*: the addition, in any amount, of an oil of different botanical origin, or of an olive oil of lower grade, to genuine and authentic olive oil (a recent example is the adulteration of olive oil with hazelnut oil)
3. *counterfeit or imitation*: whole substitution of seed oils or lower grade oils for olive oils
4. *trading fraud*: illegal practices involving the marketing and selling of olive oils in a lower amount than that defined on the label

Other important issues such as varietal or geographical origin, method of agronomical production (harvesting by hand-picking versus mechanical shaker), system of oil extraction (stone mill versus hammer crusher), come under the banner of fraud.

2.6 Official analysis methods

Olive oils must satisfy various instrumental analytical parameters and conform to legal limits, restrictions and precise panel test scores.

Table 2.3 presents a summary of the Annexes to EEC regulation 2568/91 and successive additions and amendments. Additional information has been inserted into the original summary format with indications of the purpose and scope of the content of each Annex. Annex I is a table gathering all the characteristic qualitative and quantitative data of the nine olive oil categories.

2.7 Quality parameters

A top quality olive oil must maintain its appreciated qualities throughout its lifetime, meaning that the natural, original chemical composition must remain unchanged for as long as possible, from the mill to the consumer. By 'natural original composition' of an olive oil we generally mean the complex mixture of substances present in an olive oil at the end of a particular extraction process. Considering that triglycerides and long chain fatty acids are tasteless and odourless, the organoleptic characteristics of an oil must be attributed to three groups of substances:

1. *colour* to chlorophyll and carotenoid pigments
2. *bitterness* to, in the main, phenolics
3. *attributes* such as leafy, fruity and lawn to short carbon-chain volatile compounds.

These last compounds, not present in the olive fruit, form through the lipoxygenase oxidation pathway during the extraction process, an action quite distinct from the autoxidation and photoxidation that occur during the storage of oil and that, instead, produce mainly compounds with disagreeable flavours and odours.

Official quality parameters comprise acidity, peroxide value, halogenated solvents, ultraviolet adsorption and sensory assessment. In addition to these official parameters important roles and high values are attributed to the content of the chlorophyll and carotene pigments and phenol antioxidants, as well as to the correlated induction time value. The significance of olive oil quality parameters is explained in Table 2.1. The phenol content and the related induction period, defined as 'the delay in the commencement of oxidation in an oil', merits particular comment: the longer the induction period, the better the oil. Resistance

Table 2.3 Summary of the methods of analysis to regulation EC 2568/91 and successive additions and amendments with the indication of title, purpose and scope

Title	Purpose	Scope
Annex I		
Characteristics of olive oil	To define relevant identity chemico-physical and organoleptic characteristics of olive oil	An olive oil is to be rejected if any one of the characteristics lies outside the limit laid down
Annex II		
Determination of free fatty acids	Determination of free acids in olive oils containing low concentration of these acids apart from virgin lampante and crude olive-residue oils	To assess the olive oils quality To detect hydrolysis of triglycerides, a main route for the deterioration of oils High quality oils have low free fatty acids content, but the reverse is not always true. Most customers still associate, wrongly, low acid value with high oil quality.
Annex III		
Determination of peroxide value (PV)	Determination of peroxides and hydroperoxides formed in olive oil by autoxidation and photoxidation	To determine the extent and progression of oxidation To evaluate the oil quality. To evaluate the oil's resistance to oxidation
Annex IV		
Determination of wax content by capillary column gas-liquid chromatography	Quantitative and qualitative determination of long chain aliphatic esters. The result is expressed in mg/kg	Wax esters (long chain aliphatic esters) represent a class of compounds suitable to evaluate the authenticity of olive oil. The presence of wax esters in amount greater than the official limits indicate the presence of solvent-extracted oil (olive-residue oil)
Annex V		
Determination of the composition and content of sterols	The method describes a procedure for determining the individual and total sterol content of olive oil. 'Apparent' β-sitosterol is the sum of $\Delta^{5,23}$-stigmastadienol + clerosterol + β-sitosterol + sitostanol + Δ^{5}-avenasterol + $\Delta^{5,24}$-stigmastadienol ($\geq$93%)	To uncover sophistication with oils different from olive oil. Sophistication is detected for: cholesterol >0.5%; brassicasterol $\geq$0.1%; $\geq$0.2% for olive-residue oils; campesterol >4.0%; stigmasterol > campesterol; Δ^{7}-stigmastenol >0.5%
Annex VI		
Determination of erythrodiol and uvaol	Erythrodiol (commonly understood as the diols erythrodiol and uvaol together) is a pentacyclic triterpene. The result is expressed as the percentage of diols in a mixture of diols and sterols	Erythrodiol and uvaol are found mainly on the surface of olive fruit and also on the grape seed. Their presence in amount greater than the official limits demonstrates the mixture with solvent extracted olive oil (olive-residue oil)

Table 2.3 (continued)

Title	Purpose	Scope
Annex VII		
Determination of fatty acids in the 2-position of the triglycerides	Determination of the composition of fatty acids esterified at the 2-position of glycerol. The result is expressed as the percentage of palmitic and stearic acids on the 2-position of glycerol	To uncover sophistication carried out by mixing inter-esterified oils with olive oils
Annex VIII		
Determination of the content of trilinolein	The method is a high performance liquid chromatography (HPLC). In LC triglycerides are eluted according to equivalent carbon number (ECN)	To determine the glycerides comprising a vegetable oil qualitatively and quantitatively
Annex IX		
Spectrophotometric investigation in the ultraviolet	The method describes the procedure for performing a spectrophotometric examination of oils in the ultraviolet. Absorptions at 232 and 270 nm are expressed as specific extinction $E_{1\text{cm}}^{1\%}$ conventionally indicated K. Absorptions are due to the presence of conjugated polyene systems	Spectrophotometric UV analysis of olive oil provides information of its state of oxidation occurring during conservation and also changes brought about by technological processes
Annex X		
Analysis by gas chromatography of methyl esters of fatty acids	The method describes procedures for the determination of fatty acids of an oil. The choice of the procedure is dictated by the acidity of the oil	Sophistication with different oils is proved for: myristic acid $>0.05\%$; linolenic acid $>0.9\%$; arachic acid $>0.6\%$; eicosenoic $>0.4\%$; behenic $>0.2\%$ for oils 1–6 and $>0.3\%$ for oil types 7–9; lignoceric $>0.2\%$
Annex XI		
Determination of volatile halogenated solvents content of olive oil	The method is head space gas chromatography. The result is expressed as mg/kg	Olive oil can be contaminated by halogenated solvents during storage, during production or by accident
Annex XII		
Organoleptic assessment of virgin olive oil	The purpose of this method is to determine the criteria needed to assess the flavour characteristics of virgin olive oil and to develop the methodology required to do so	The method described is only applicable to the organoleptic assessment and classification of virgin olive oil. It confines itself to grading the virgin oil on a numerical scale related to the perception of its flavour stimuli, according to the judgement of a group of selected tasters working as a panel

Table 2.3 (continued)

Title	Purpose	Scope
Annex XIII		
Proof that refining has taken place	Neutralization and decoloration of olive oil in the laboratory	To prepare the oil sample for additional analysis, e.g. ultraviolet
Annex XIV		
Additional note 2, 3 and 4 to Chapter 15 of the combined nomenclature		
Annex XV		
Oil content of olive residue	Self-explanatory	Self-explanatory
Annex XVI		
Determination of iodine value	A method for the determination of iodine value of olive oil. The iodine value is expressed as g of iodine per 100 g of oil	To obtain information on degree of unsaturation of an oil
Annex XVII		
Determination of stigmastadienes in vegetable oils	Determination of stigmastadienes in vegetable oils containing low concentration of these hydrocarbons, particularly in virgin olive and crude olive-residue oil	The standard may be applied to all vegetable oils although measurements are only where the content of these hydrocarbons lies between 0.01 and 4.0 mg/kg. The method is particularly suited to detecting the presence of refined vegetable oils (olive, olive residue, sunflower, palm, etc.) in virgin olive oil since refined oils contained stigmastadienes and virgin oils do not
Annex XVIII		
Determination of the difference ECN42 HPLC – ECN42 calculated	Quantitative determination by HPLC of triglycerides in olive oil with ECN42. Determination of theoretical triglicerides with ECN42 calculated on the basis of percent concentration of fatty acids according to the 1,3-random-2-random pattern	In olive oils, the difference ECN42 the HPLC – ECN42 theoretically lies within narrow limits. When the analysis of an oil sample gives results that are outside the limits laid down, that is understood as a proof of sophistication with seed oils

ECN, equivalent carbon number.

to oxidation (autoxidation, photoxidation, or their combination) depends on both chemical factors (accessibility to oxygen, degree of unsaturation, metal compounds, antioxidants such as phenols) and physical factors (temperature and light). It is obvious that the physical factors depend greatly on technology and household practices.

2.8 Chemical and chemico-physical analysis

Official analytical methods provide data and results that give sound proof of the authenticity and quality of an olive oil, or, alternatively, elements that permit the uncovering of altered and sophisticated 'over-classified' oils. This section is not to be regarded as a comprehensive presentation of olive oil chemistry and analysis but is intended only to give the reader an overall perspective of the origin and fate, and the chemico-physical properties, of the classes of compounds constituting olive oil.

Wherever it is considered useful there is the development of simple analytical and structural chemistry elements, as well as of reaction mechanisms, to explain the chemical properties and the induced or spontaneous transformation of the natural oil products. We do not pretend that the official methods presented, and commented on, in this section are the most important for the determination of olive oil authenticity, but we do hope that the methodologies illustrated will represent a foundation upon which the interested reader will achieve familiarity with some of the original classes of compounds and those of new formation that make up olive oil.

For each method considered we highlight, and comment on, the major functions:

1. detection of alteration and sophistication and possible identification of undeclared seed oil or lower grade olive oil
2. measurement of quality parameters.

2.9 Oxidation

2.9.1 Lipid hydroperoxides

Lipid hydroperoxides are the compounds formed when atmospheric oxygen enters the lipid molecular moiety, producing the ROOH compounds. There are three oxidation mechanisms giving rise to hydroperoxides.

2.9.2 Autoxidation

Autoxidation occurs when reactive free radical lipids interact with triplet diradical oxygen in its ground state, according to the mechanism in Figure 2.2. In the case of the oleates, the autoxidation reaction produces a mixture of allylic hydroperoxides substituted at positions 8, 9, 10 and 11, with the double bond remaining at the original position (Δ9–10) or appearing at the two adjacent C–C bonds Δ8–9 and Δ10–11 (Figure 2.3) (Porter *et al.*, 1995).

In the case of linoleate, autoxidation has been shown to produce the hydroperoxides shown in the scheme with the formation of conjugated systems absorbing in the ultraviolet region. Linoleate regioisomers are formed only at positions 9

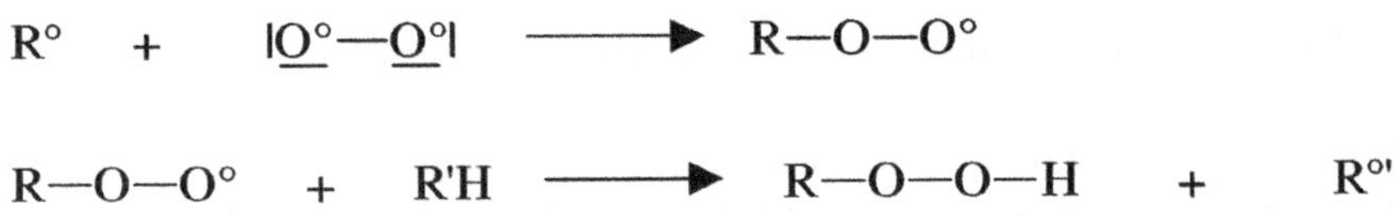

Figure 2.2 Lipid hydroperoxides from autoxidation.

and 13 but, allowing for the different configurations of the double bond, the number of isomers totals four (Figure 2.4).

2.9.3 Photoxidation

When ground-state triplet oxygen is excited into a higher energy state by energy transfer from a sensitizer such as chlorophyll, singlet oxygen, where all the electrons are paired, is formed, probably entering the so called 'ene' pericyclic reaction according to the mechanism shown in Figure 2.5. Singlet oxygen reacts with linoleates forming the six isomers shown in Figure 2.6. Four isomers out of six contain conjugated diene systems and an isolated double bond (Frankel, 1991).

2.9.4 Lipoxygenase oxidation

Molecular oxygen, as distinct from reactions involving radicals or singlet oxygen, is directly inserted into free fatty acids by lipoxygenase (LOX) enzymes. Lipoxygenases, both regio- and stereospecific enzymes, react on the 1,4-pentadienyl moieties such as those of linoleic and α-linolenic acids.

Lipoxygenases play an important role in determining oil quality. The aromas typical of olive oil are due to complex mixtures of volatile compounds; those especially abundant are saturated and unsaturated six carbon atom aldehydes, alcohols, and the esters of alcohols, the cited aldehydes and alcohols having sensory properties responsible for the so-called green odour (Hatanaka, 1996).

All these volatile substances are synthesized through a series of reactions collectively called 'the lipoxygenase pathway' shown in Figure 2.7. Whilst both 9- and 13-hydroperoxides are formed by lipoxygenase, the lyase cleaves the 13-hydroperoxides but does not act on 9-hydroperoxides. A major component of the volatiles is the aldehyde 2-*trans*-hexenal (Olias *et al.*, 1993; Salas *et al.*, 1999).

2.9.5 Transformation of hydroperoxides

Hydroperoxides are relatively non-volatile compounds that break down even at room temperature. The hydroperoxide can undergo homolysis to form an

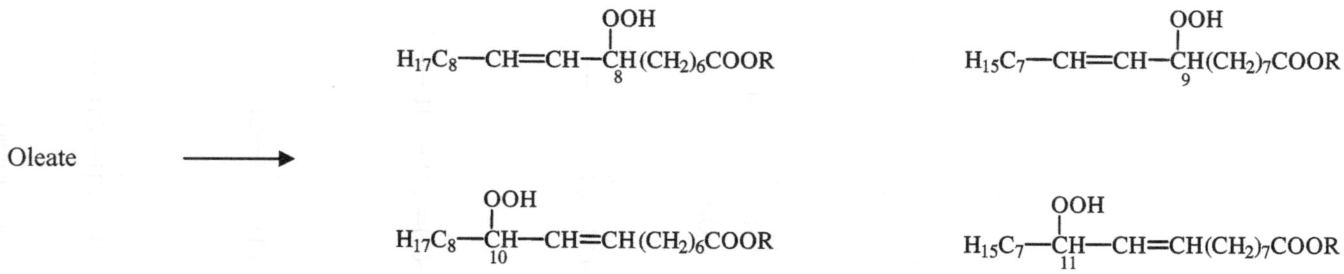

Figure 2.3 Regiomeric hydroperoxides from autoxidation of oleates.

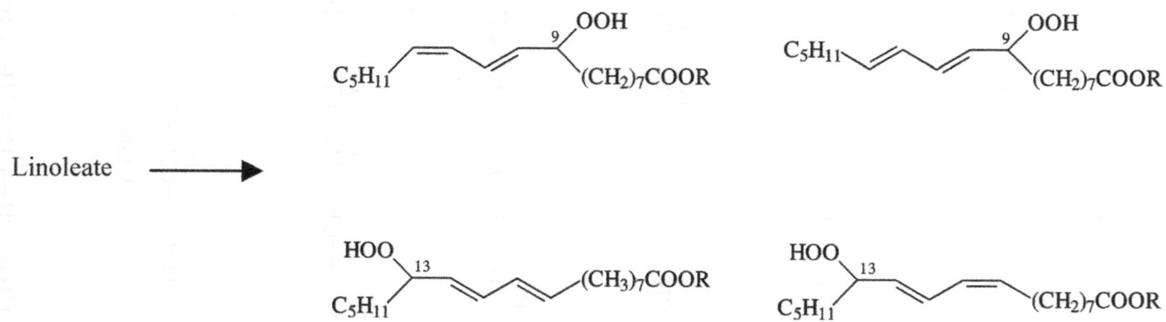

Figure 2.4 Major hydroperoxide isomers from autoxidation of linoleates.

Figure 2.5 'Ene' pericyclic reaction by photoxidation.

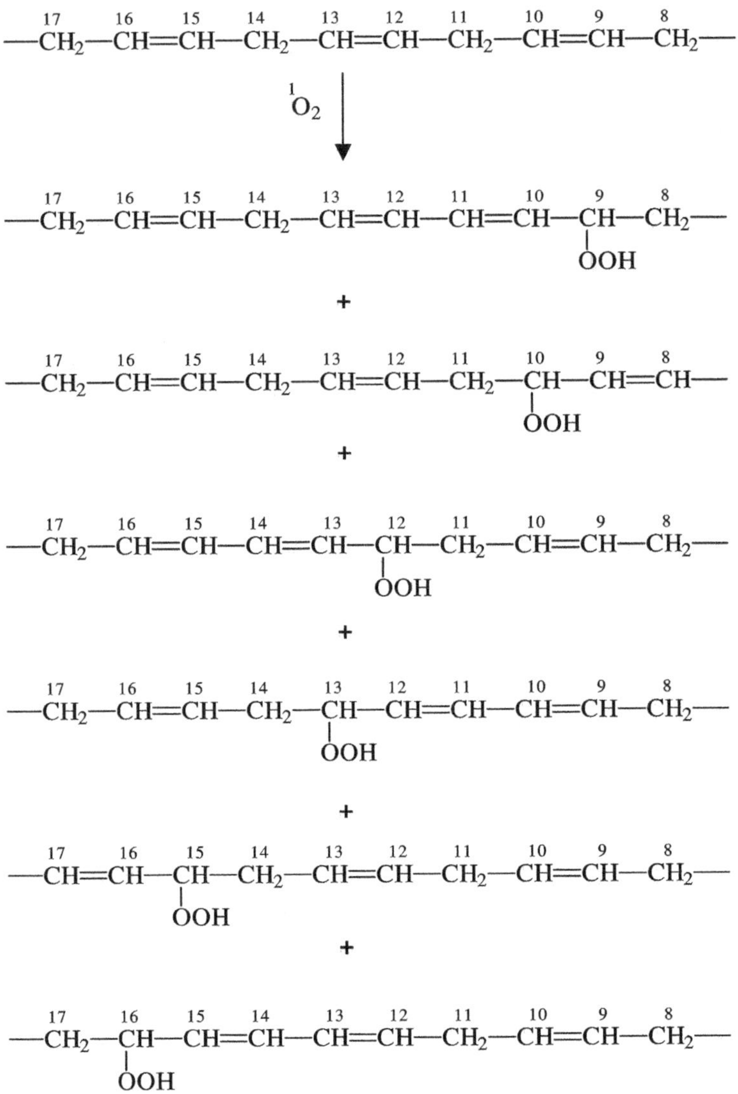

Figure 2.6 Regiomeric hydroperoxide by photoxidation of linoleates.

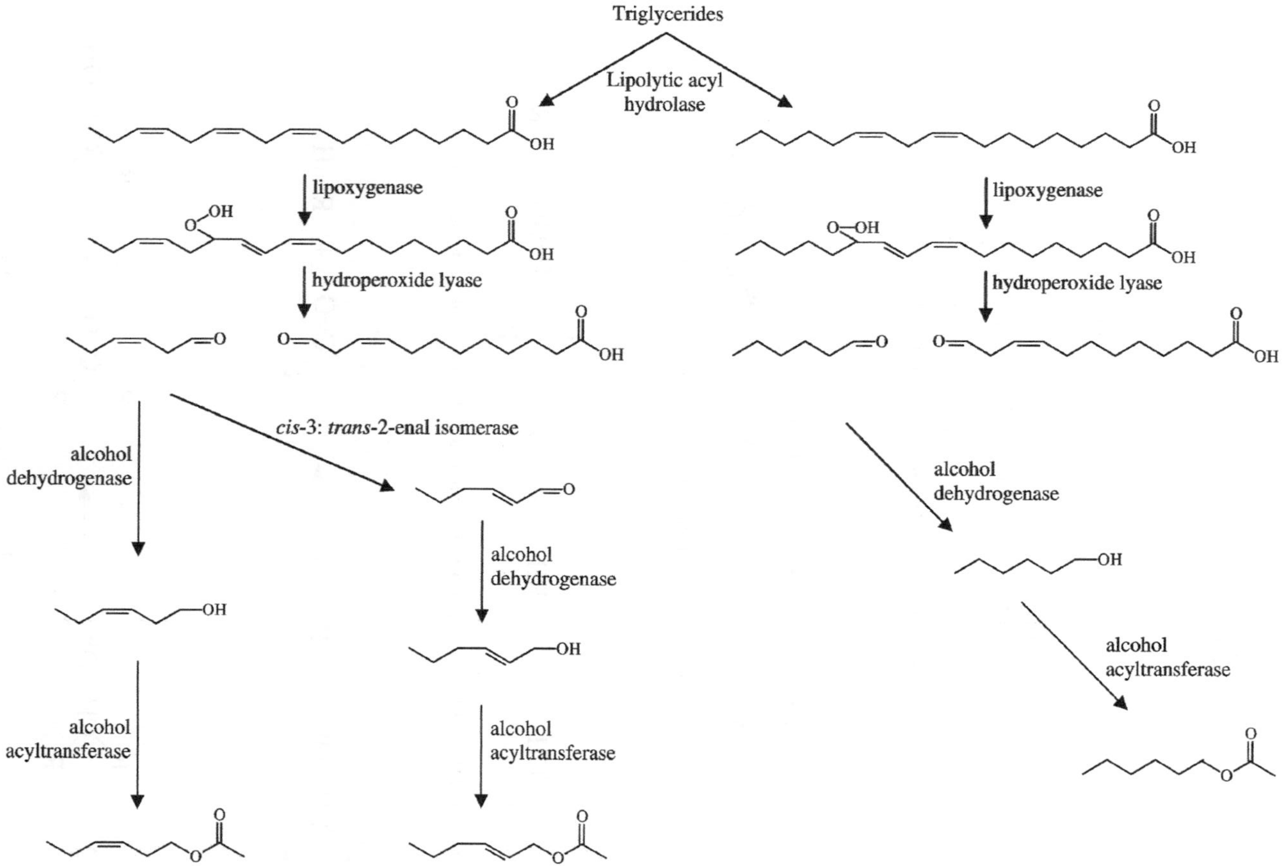

Figure 2.7 The lypoxygenase pathway.

alkoxy radical, an intermediate reaction leading to a great number of compounds (Hamilton *et al.*, 1997). The alkoxy radical arising from homolysis undergoes fragmentation, termed β scission, in which either the β C–C bond to oxygen is broken, giving an aldehyde and a radical, or, alternatively, reduced, producing an alcohol as shown in the equation (Figure 2.8).

Alternative major reactions that alkoxy radicals can undergo are:

1. abstraction of a hydrogen atom producing an alcohol and a new radical (Figure 2.9)
2. disproportionation with another radical giving rise to a carbonyl compound, alcohol and a reduced radical (Figure 2.10).

The sequence of events in which triglycerides and free fatty acids or derivatives are involved in the oxidation process has been studied by many researchers. The

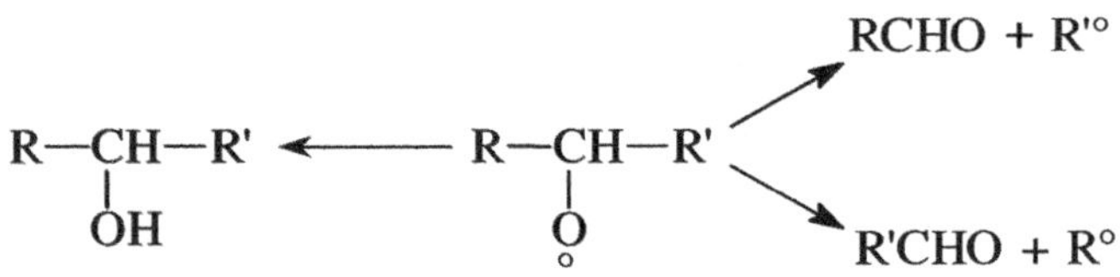

Figure 2.8 Reduction and β scission of alkoxy radical.

Figure 2.9 Abstraction of a hydrogen atom by an alkoxy radical from a susceptible molecule.

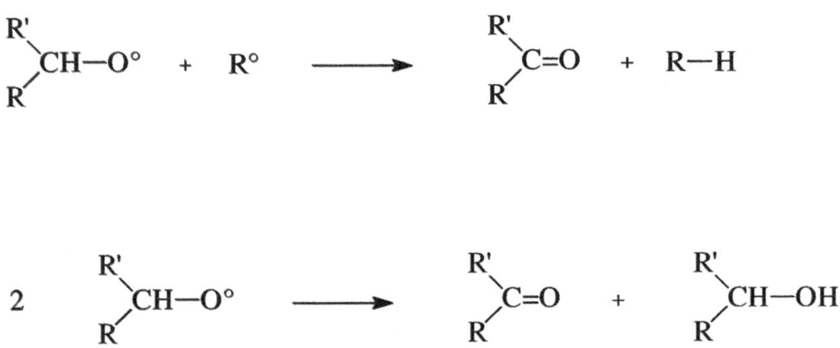

Figure 2.10 Alkoxy radical disproportionation.

most significant reaction steps involving a triglyceride are shown in the scheme proposed by Hoffmann (1970) (Figure 2.11).

The first isolable product of oxidation of the triglyceride is a tasteless and odourless hydroperoxide. The hydroperoxide can undergo homolysis, usually catalysed by a metal ion, to form an alkoxy radical whose decomposition may follow either β or β′ scission. β Scission gives a free aldehyde with flavour and/or odour along with an intermediate free triglyceride radical that can react further. Alternatively the alkoxy radical decomposes according to β′ scission, leading to a bound, odourless aldehyde and an R° radical that can react with other neutral molecules, or with itself, to form R–R.

As a further example: the four hydroperoxides obtained in the autoxidation of oleate would be expected to give either the aldehydes and radical esters shown in the following equation or, alternatively, the ω-oxoesters and alkane and alkene radicals if the β scission takes place on the other C–C bond. The free radicals can then react with neutral molecules or inactivate one another (Figure 2.12).

As already stated, hydroperoxides can enter a reduction reaction leading to alcohols that can, in turn, give rise to conjugated polyenes by elimination of parts of a water molecule as shown in the equation in Figure 2.13.

2.9.6 *Ultraviolet absorption to detect oxidation and refining*

The ultraviolet (UV) region of near UV covers the 200–400 nm wavelengths. On considering the chemical functional groups of all the substances originally present in the olive fruit it can be seen how only the high energy region around 200–210 nm is suitable for interaction with the carboxylic groups and isolated double bonds of the molecular species of olive oil. The concentration of aromatic components like simple phenols and phenyl alkyl esters is too low to absorb UV radiation at a detectable level. Thus, the UV spectrum of a genuine, properly conserved, olive oil will show practically no absorption peaks.

Instead, when there are peaks, indicating UV absorption, this is a clear indication that the oil has undergone either: 1. an oxidation process; or 2. a thermal deodorizing or bleaching treatment in the presence of active earth during the refining processes.

2.9.7 *Ultraviolet absorption* K_{232}, K_{270} *and* ΔK

The UV method is based on measurements of extinction or optical density in the 232 nm and 270 nm regions in which diene and triene conjugated systems are known to absorb UV. The intensity of the UV spectral absorption band at 232 nm and 270 nm is also useful to detect, and also in part to quantify, both the extent of the oxidation of monounsaturated and polyunsaturated acid moieties

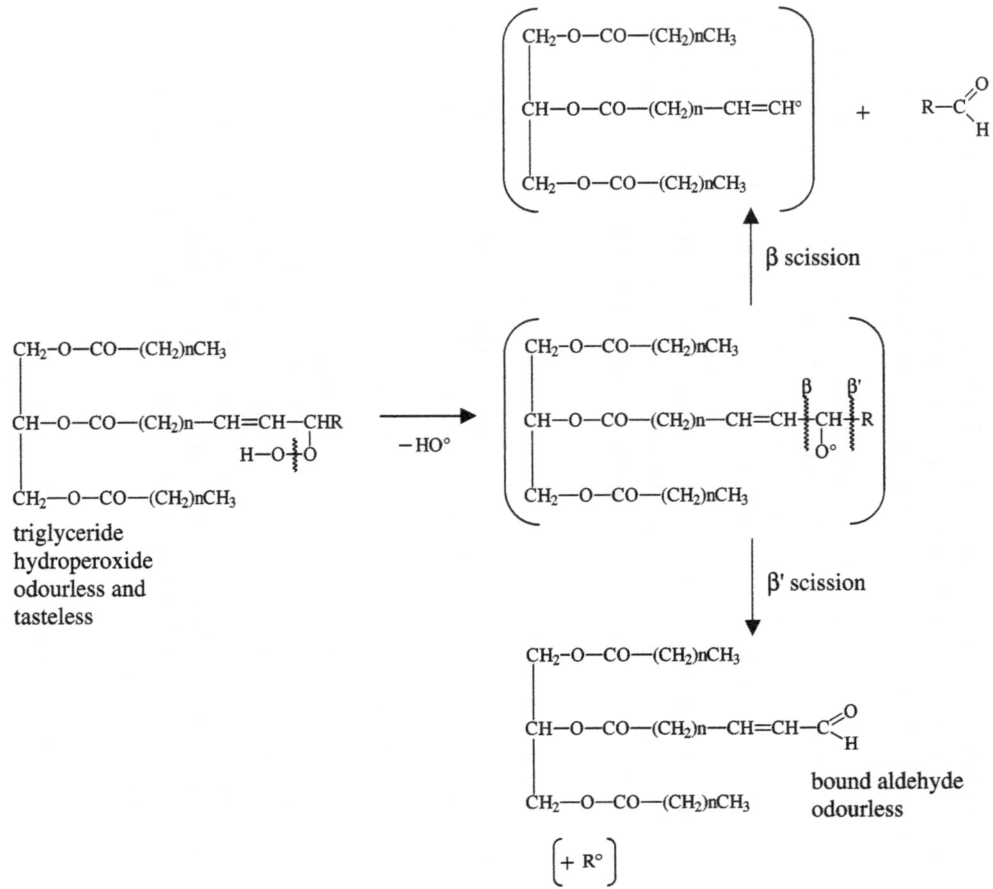

Figure 2.11 Hoffmann's oxidation scheme of triglycerides.

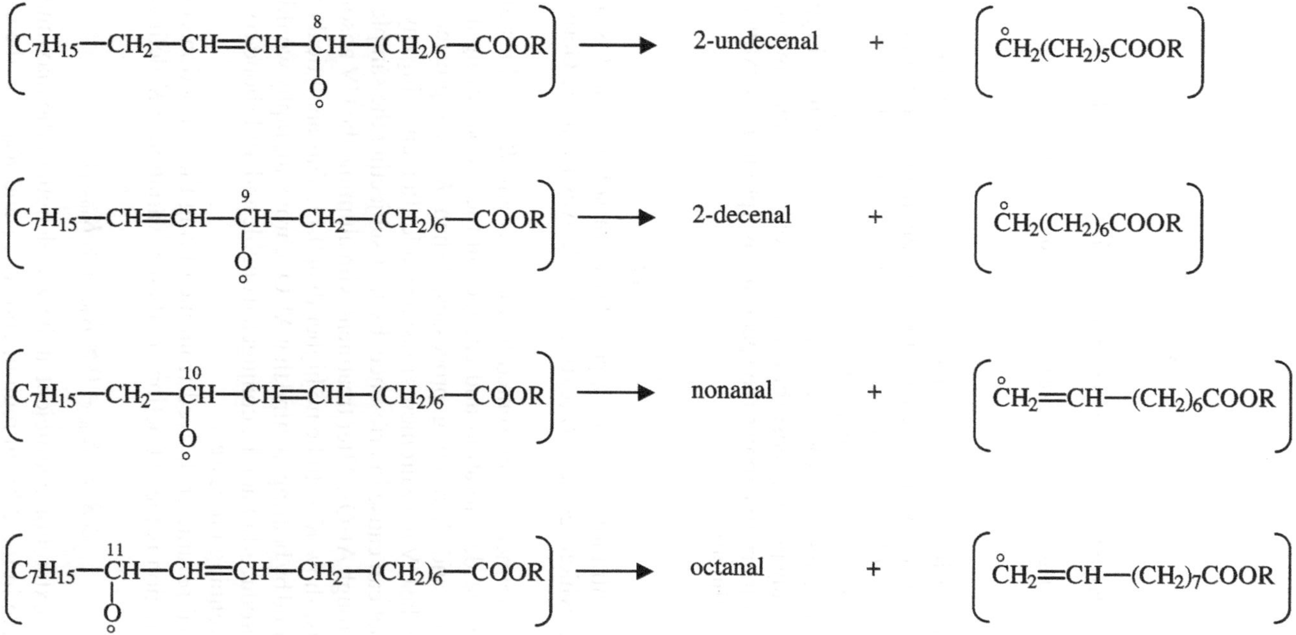

Figure 2.12 Decomposition of alkoxy radicals to give aldehydes.

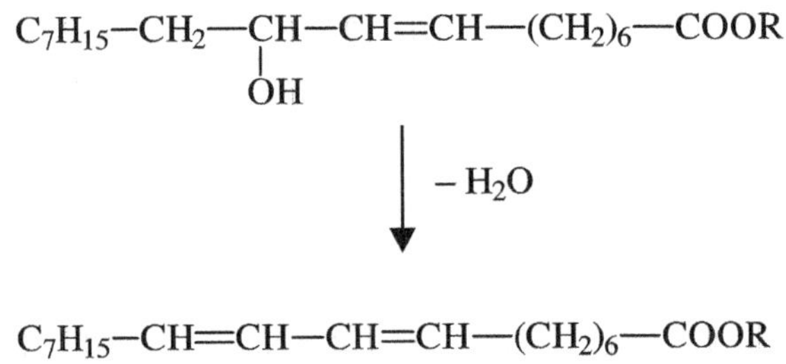

Figure 2.13 Conjugated diene from hydroperoxides.

and their decomposition products. The method takes no account of the UV region near 310–320 nm where there occurs maximum absorption of conjugated tetraenes that have seldom been documented for olive oil. Typical molecular structures that absorb UV light in the 232 and 270 nm regions, and whose formation was discussed in the previous section, are shown in Figure 2.14. Because of the many possible absorbing oil components of unknown molecular weight, it is customary to express the spectral absorption, E as extinction, one per cent, one centimetre as

$$K = E^{1\%}_{1\,\mathrm{cm}} = \frac{E}{cl} \tag{2.1}$$

where c is concentration of oil in g per 100 ml of solvent and l is in cm. In this formula K, which replaces E (extinction), derives from the German word *Konjugation*.

The quality parameters represented by the K_{232} and K_{270} values give indications of possible deterioration and changes in the oil occurring as a result of: 1. oxidation; and 2. refining processes. When K_{270} is greater than the standard value, the UV measurements must be made after removing any oxidized components that can cause interference, by first subjecting the sample to chromatography through Al_2O_3. After treatment with alumina, the UV absorption of the sample must show K_{270} to be no higher than 0.11. This analytical operation can be understood by bearing in mind that Al_2O_3 chromatography would restrain the oxidation products but not the conjugated double and triple bond compounds formed in the refining processes.

An additional parameter derived from measuring the extinction coefficient in the 270 nm region is the extinction coefficient variation ΔK defined by the equation

$$\Delta K = K_{\mathrm{m}} - 0.5(K_{\mathrm{m}-4} + K_{\mathrm{m}+4}) \tag{2.2}$$

in which K_{m} = extinction coefficient at the wavelength of the maximum of the absorption curve in the 270 nm region; and $K_{\mathrm{m}-4}$ and $K_{\mathrm{m}+4}$ = the extinction coefficients at wavelength 4 nm lower and higher than the K_{m} wavelength.

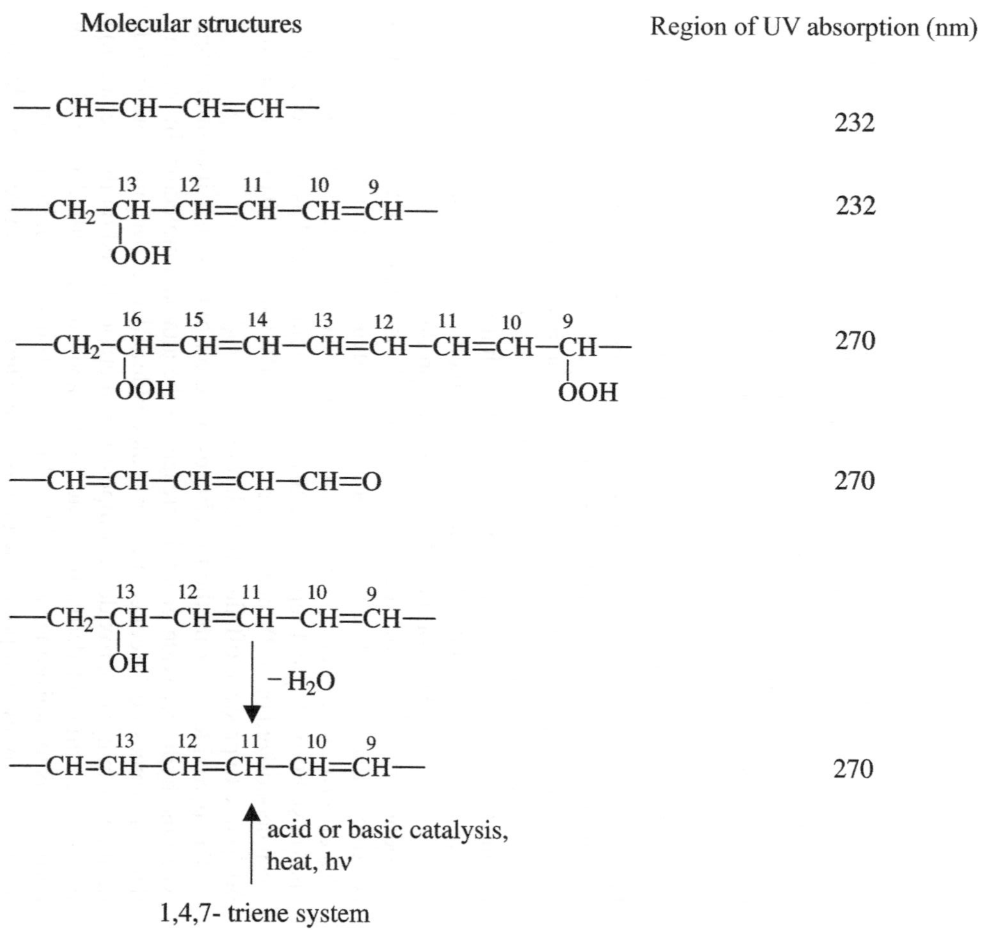

Figure 2.14 Absorption of ultraviolet by selected molecular structures.

Measurements of the first three categories of virgin olive oil must give a ΔK value not higher than 0.01: for refined olive oil and olive oil the limits are set at 0.16 and 0.13 values, respectively, whilst for oils of types 8 and 9 ΔK should not exceed the values 0.25 and 0.20, respectively.

2.9.8 *Double-bond migration to give conjugated polyenes*

Under drastic chemical and thermal conditions the double bond of many unsaturated compounds is shifted. If there is the possibility of conjugating with a double bond already present in the molecule, it goes towards it to make a more thermodynamically stable polyene (Smith and March, 2001).

In the case of olive oil the original substrate that can possibly undergo double-bond migration is that of linoleate and linolenate. The hypothetical reaction mechanism involves: (a) a base, (b) an acid and (c) a radical (Figure 2.15).

2.9.9 *Peroxide value, anisidine value and thiobarbituric acid test*

From the above description of a molecular species absorbing in the UV wavelength range, it appears that the UV test is not wholly specific for substances produced in lipid peroxidation. Therefore other methods are needed to detect and evaluate lipid oxidation. Among the variety of methods available in the literature, iodometry is the chosen official method, although it fails when hydroperoxides are present in low amounts. Note also that iodometry will measure the peroxides present in the oil, but not their decomposition products.

Iodometry on peroxidized olive oils gives the peroxide value (PV), which is expressed as the milliequivalent of active oxygen in the peroxides present in 1 kg of olive oil. Experienced oil analysts can relate PV values with rancidity and oil stability. Typical reports are that a fresh genuine oil has a PV of 2–5, a fresh deodorized oil a PV near to zero and that PV values of 10 or more are for oils prone to produce off-flavour volatiles.

The peroxide value method evaluates olive oil oxidation at its initial stage when the hydroperoxide group is present in the attacked molecule. In order to detect and measure the concentration of aldehyde and other carbonyl compounds derived from hydroperoxide decomposition, methods based on the preparation of suitable carbonyl derivatives had to be devised. A method of evaluating carbonyls is the anisidine value (AV); this consists in derivatizing the aldehydes and other carbonyl compounds with anisidine. The reaction between an aldehyde and the amino group gives rise to an anil with a chromophoric group that absorbs at around 350 nm. An alternative, but unofficial, test used by many investigators is that of thiobarbituric acid (TBA) (Rossell, 1989). This test is explained as being due to the reaction of malondialdehyde with TBA as shown in the equation (Figure 2.16). Malondialdehyde is formed in very low amounts from a fatty chain with three or more double bonds (Frankel, 1991).

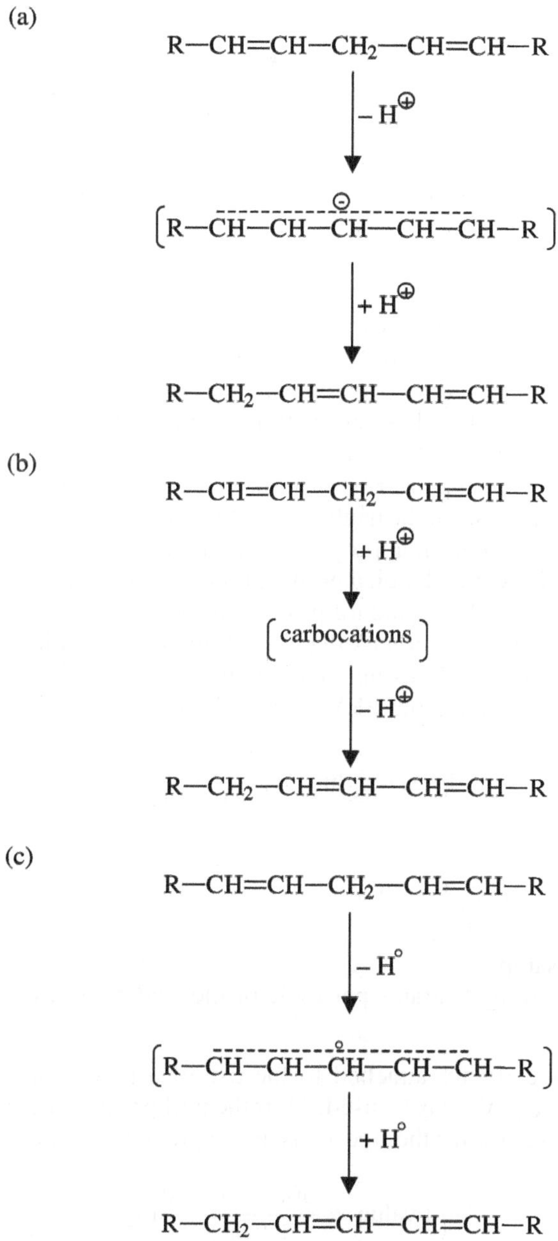

Figure 2.15 Double-bond migration in 1,4-diene systems.

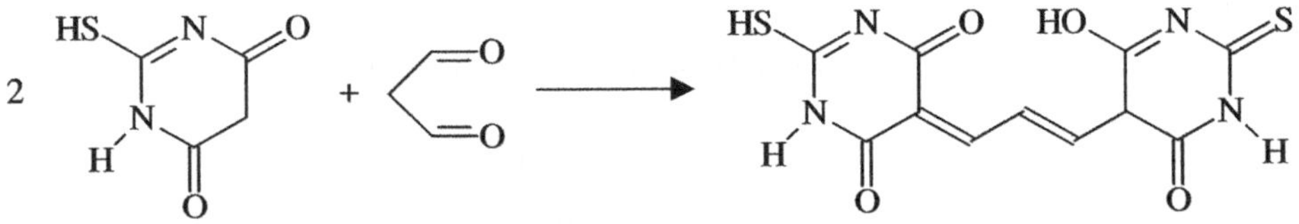

Figure 2.16 Reaction of malondealdehyde with thiobarbituric acid.

2.10 Free fatty acids

Free fatty acids (FFAs) are always found, even in genuine extra virgin olive oil and, to the best of our knowledge, there have been no reports of FFAs being present in amounts lower than 0.2–0.3%, not even in the very best fresh extra virgin olive oil. Thus these original free acids remain from triacylglycerol synthesis or are, alternatively, produced separately from triglyceride synthesis.

However, when measuring FFAs in olive oil, consideration is given to the product of the hydrolysis of the triglycerides by free lipolytic enzymes, microorganisms or simply water, in appropriate pH and temperature conditions. Thus the formation of FFAs in oil indicates that the commodity has suffered abnormal damaging conditions. The standard procedure for determining such acids is to dissolve a sample in 50:50 ether–ethanol and titrate the solution with ethanolic potassium hydroxide 0.1 M using phenolphthalein as an indicator.

The formula for calculating FFAs is the following:

$$\text{FFAs(acidity)} = \frac{V \times c \times M}{10 \times m} \tag{2.3}$$

where:

V = ml of KOH solution,
c = normality of KOH,
m = grams of sample,
M = molecular weight, grams per mole of the acid chosen for expression of the results.

To avoid any uncertainty attached to the use of a particular acid molecular weight, acid value (AV) may be used; AV is the milligrams of potassium hydroxide necessary to neutralize the free acids in one gram of oil, and is calculated as:

$$\text{Acid value} = \frac{56.1 \times V \times c}{m}\,\text{mg/g} \tag{2.4}$$

where:

V = ml of NaOH or KOH solution used
c = concentration in moles per litre of NaOH or KOH solution
m = grams of oil

The usefulness of the test, as such, in order to assess the grade and the quality of an oil, is doubtful. A low, free acidity value makes sense when comparing oils that have not undergone any refining process. In fact, from Table 2.2 it can be seen that all the oils in categories 5 to 9 have acidity percentage values equal or less than that of virgin olive oil. Furthermore, refined olive oil, which has undergone alkali treatment, has an upper limit as low as 0.5%.

The usefulness of the test would be enhanced if it were possible to determine the precise chemical nature of the free acids. For instance, by determining the free acids originally present, we should be able to obtain further useful information about both the biosynthesis, and the degradation, of the triacylglycerols.

2.11 Fatty acid composition

Fatty acid determination in oils is extremely important, enabling lipid scientists to classify an oil correctly, both botanically and in accordance with commercial gradings. The two most common reactions for preparing the methyl esters of fatty acids, both free and esterified as triglycerides, are:

1. direct *trans*-esterification in anhydrous methanol in the presence of an acidic catalyst (sulphuric acid, hydrogen chloride, boron trifluoride) or basic catalyst (potassium hydroxide, sodium methoxide).
2. esterification proceeding through a first step in which the oil sample is heated in methanol-sodium methylate, followed by a second step in which the reaction mixture of step one is heated in the presence of concentrated sulphuric acid to give the methyl esters of the originally free fatty acids, together with the methyl esters of acids of glycerides.

The procedure of method 1. is quicker than method 2. and will give a complete picture of the fatty acid composition of an oil only when no free acids are present in the sample; method 2. needs to proceed through two steps but produces the most complete picture of both the free and esterified fatty acids comprising an olive oil.

The methyl esters obtained are readily analysed qualitatively and quantitatively by gas chromatography, and the data obtained allow detection of sophistication. In the literature there is a wealth of fatty acid analysis data on virgin olive oils, all constantly reporting almost the same qualitative composition. However, what is surprising is the systematic attitude of so many researchers not reporting the presence of vaccenic acid, 11-*cis*-octadecenoic acid, an isomer of oleic acid. The presence of this positional isomer of oleic acids was first described in olive oil by Tulloch and Craig (1964).

2.11.1 Detecting seed oils

1. presence of lauric acid: oil babassu, coconut, palm kernel
2. presence of myristic acid >0.05%: oil babassu, coconut, palm kernel

3. presence of palmitic acid >20%: palm
4. presence of stearic acid >5%: peanut, soyabean, sunflower
5. presence of oleic acid <55%: babassu, coconut, cottonseed, grapeseed, maize, palm kernel, safflower
6. presence of linoleic acid >21%: peanut, cottonseed, grapeseed, maize, safflower, sesame, soyabean, sunflower
7. presence of linolenic acid >0.9%: rapeseed, soyabean
8. presence of arachidic acid >0.6%: peanut
9. presence of eicosenoic acid >0.4%: rapeseed
10. presence of behenic acid >0.2%: peanut, sunflower
11. presence of erucic acid: rapeseed, safflower
12. presence of lignoceric acid >0.2%: peanut, rapeseed.

2.11.2 Trans *fatty acids in refined and deodorized oils*

Mono- and polyunsaturated fatty acids occurring in olive oil have the sole *cis* configuration and are stable molecules. The activation energy ΔG necessary to permit the change from the *cis* to *trans* configuration is about 30 kcal/mol, and, consequently, the probability of geometrical isomerization is practically negligible in normal conditions (Figure 2.17).

However, in certain temperature conditions and in the presence of a number of catalysts, oleic acid, linoleic acid and linolenic acids are converted into *trans* isomers that comprise various combinations of *cis*, *trans* double bonds in the acid moiety. The *cis–trans* isomerization by linoleic acid during deodorization or refining of oils was studied in an experimental pilot plant using nitrogen as the stripping gas, in place of steam. Samples of bleached sunflower, olive and soybean oils were tested at temperatures in the range 240–265°C. The activation energy was higher for the formation of acid C18:2 (9t, 12c) than that for C18:2 (9c, 12t), whilst acid C18:2 (9t, 12t) showed the lowest value (Leon-Camacho *et al.*, 2001).

These *trans* compounds have different physical properties than natural *cis* isomers and are readily separated with gas chromatography analysis. The detection of *trans* isomer fatty acids in an olive oil indicates addition of seed oils to olive oil, of refined olive oil to virgin olive oil and of esterified oil to residue olive oil and olive oil.

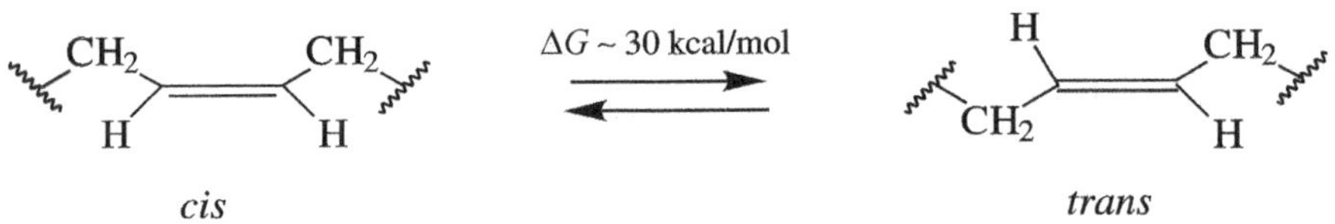

Figure 2.17 *cis–trans* Isomerization.

Gas chromatography is the most accurate means of quantifying the *trans* isomers present in fatty acid mixtures. In the GC trace the *trans* peaks appear broader because they contain isomers whose complete resolution is difficult. Virgin olive oils may show, in the sum of the *trans*-oleic acid-isomers, a maximum value of 0.05% and in the sum of *trans*-linoleic and linolenic acid the same maximum value of 0.05%. Lower grade olive oils and refined olive oils are allowed wider limits for fatty acid *trans* isomers (see Table 2.2).

A number of alternative methods are available for determining *trans*-isomers, including spectroscopy, both infrared (IR) and Raman, and nuclear magnetic resonance (NMR) (Firestone and Sheppard, 1992).

2.12 High performance liquid chromatography criteria for detecting sophistication with seed oils

The following elementary liquid chromatography (LC) elements could help to understand the method. In LC, triglycerides are separated according to equivalent carbon number (ECN). Elution order is determined by calculating ECN, normally defined as CN-2n, where CN is carbon number and n is the number of double bonds present in the triglycerides (Firestone, 1994).

Despite the vast range of high performance liquid chromatography (HPLC) analytical data for olive oils available in the literature, there have been no proposals on the part of any official authority for defining triglyceride ranges to classify the various grades of olive oil or to distinguish such oils from other oils.

In contrast with the fatty acids, where the typical composition permits the uncovering of several types of sophistication, triglyceride molecular species do not represent straightforward data capable of discriminating olive oil from other oils.

Flor *et al.* (1993) were the first to develop criteria for the authentication of olive oil based on vegetable oil HPLC data. They observed that corn, cottonseed, soyabean, sunflower and safflower oils, to mention the most important commercial products, have large peaks for LLL, LLO and LLP but generally smaller LOO and LOP peaks (abbreviations: P, palmitic; O, oleic; S, stearic; L, linoleic; Ln, linolenic: Po, palmitoleic). Additional typical peaks were observed: LnLL peak (ca. 7%) in soyabean and LnLO peak (ca. 7%) in rapeseed oils, respectively. Other relevant compositional pictures were observed: peanut oil displays a relatively small LLL peak (ca. 3.5%) but larger LLO and LLP peaks (ca. 18.2, 5.9%, respectively).

In olive oil, in general, the largest triglyceride peaks are LOO, LOP, OOO, POO, POP and SOO. After examining various arithmetic operations on the peak areas with the aim of setting ranges for identifying the various grades of olive oils to distinguish and detect sophistication with other oils, Flor *et al.*(1993) found that the correlation OOO/POO versus LOO/LOP represented a reliable criterion for classifying oil samples. According to the results obtained from a study on 99 oil samples, adherence to this correlation line was compulsory for virgin olive oils: oil samples not on the line were: (i) not olive oils at all; (ii) blends of olive oil with other vegetable oils; (iii) re-esterified oils. A further observation was that authentic olive oils have a trilinolein (LLL) peak lower than 0.5%.

The paper by Flor *et al.* (1993) was probably a stimulus for the official method in the EEC regulation 2568/91 for detecting seed oil sophistication of olive oil, defining the difference as ECN42 HPLC – ECN42.

The method consists essentially in a comparison of the experimentally determined concentration in percentage of natural triacylglycerol types with ECN42, with a calculated concentration of all the theoretically possible triacylglycerols on the basis of the major fatty acid composition of the oil under scrutiny. Considering the fatty acids of olive oil, the possible triglycerides with ECN42 are OLLn, LLL, PLLn, PoOLn, PoLL, PPoLn, PoPoL, SLnLn, PoPoPo and their possible positional isomers.

It is well known that in a naturally occurring oil the fatty acids are not distributed randomly on the glycerol skeleton. Similarly, in olive oil, the fatty acid pattern conforms to the 1,3-random-2-random distribution pattern. The saturated fatty acids are almost exclusively found at the 1,3-positions, whilst the 2-position is occupied almost entirely by the unsaturated Po and O, L, Ln acids.

Accordingly, the fatty acids will be distributed in two identical pools for the 1- and 3-positions of glycerol, and a third pool for the 2-position. The concentration of saturated fatty acids at the 2-position is derived by multiplying their total percent concentration by the coefficient 0.06 that, in general, will produce values lower than the 1.3% limit laid down in Annex I of regulation EC 2568/91 (EC Council, 1991) (Table 2.2). Once the correct concentration of all the fatty acids on the three glycerol positions has been calculated, the method proceeds to the calculation of the theoretical concentration of triacylglycerides with ECN42. Generalized methods for calculating the weight ratios or mole fractions of triglycerides in natural oils was developed by Vander Wal (1960, 1963) and Hayakawa (1967). It has been proved that the ECN42 HPLC – ECN42 calculated is suitable for detecting even very small quantities of seed oils in olive oil due to the high triglyceride concentrations, especially of LLL, in seed oils.

2.13 Analysis of sterols, sterenes, erythrodiol and uvaol

For many different vegetable oils there is a characteristic qualitative and quantitative distribution of the tetracyclic and pentacyclic triterpenes, and this is thus utilized as a fingerprint for their identification. These compounds are present in either the free or the esterified form.

The method applied consists in a prior, basic hydrolysis (saponification) of the oil sample, followed by thin layer chromatography (TLC) fractionation of the classes of compounds, namely hydrocarbons, tocopherols, long-chain aliphatic alcohols, triterpenic alcohols, methyl sterols, sterols and triterpenic dialcohols. This procedure loses all information about the combination of polycyclic triterpenes with fatty acids.

The homogeneous separated fractions, with the exception of hydrocarbons, are derivatized by silylation in order to reduce their polarity, and analysed in capillary gas chromatography.

The gas chromatography (GC) tracing allows a fast qualitative and quantitative screening of the components of the fraction of interest. Identification of the components and a percentage evaluation provides highly specific information about the expected purity of the olive oil.

2.13.1 Sterols

Sterols, free and esterified, are a relatively abundant fraction of the minor components of olive oil. The regulation requires a concentration ≥1000 mg/kg for virgin, refined and ordinary olive oil, whilst the content must be higher for the other types of olive oil (see Table 2.2, Annex I).

Out of the fifteen identified sterols comprising the fraction, the components characterizing olive oils are cholesterol (≤0.5%), brassicasterol (≤0.1–≤0.2%), campesterol (≤4.0%), stigmasterol (<campesterol), β-sitosterol 'apparent' (an old gas chromatographic definition for a packed column unresolved group of sterols) as sum of $\Delta^{5,23}$-stigmastadienol + clerosterol +β-sitosterol + sitostanol +Δ^{5}-avenasterol +$\Delta^{5,24}$-stigmastadienol (≥93%), Δ^{7}-stigmastenol (≤0.5). Brassicasterol, virtually absent in olive oil, is a component of some seed sterols; campesterol, stigmasterol and Δ^{7}-stigmastenol are found in several seed oils in higher concentrations than in olive oil. The following concentration figures show oils that might be admixed to olive oil:

- concentration of cholesterol >0.5%: admixed oil can be sunflower, rapeseed, palm, palm kernel
- concentration of brassicasterol >0.2%: admixed oil can be rapeseed, safflower
- concentration of campesterol >4%: admixed oil can be peanut, babassu, coconut, cottonseed, soyabean, sunflower, rapeseed, safflower, sesame, grapeseed, maize, palm, palm kernel
- concentration of stigmasterol >4%: admixed oil can be peanut, coconut, soyabean, sunflower, rapeseed, sesame, grapeseed, palm, palm kernel
- concentration of Δ^{7}-stigmastenol >0.5%: admixed oil can be sunflower, safflower.

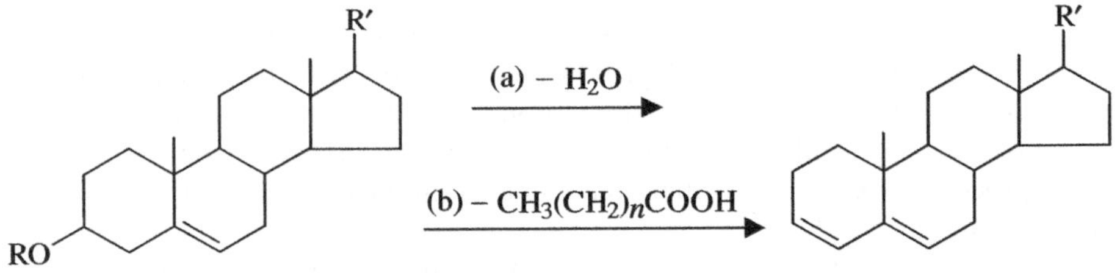

R = (a) H; (b) $CO(CH_2)_nCH_3$

Figure 2.18 Formation of steradienes.

2.13.2 Sterenes

In accordance with chemical nomenclature the sterenes can be considered a subclass of steroids as they share with these the hydrogenated cyclopentanophenanthrene carbon skeleton. Sterenes are not naturally occurring substances, but are artefacts arising from free or esterified sterols through the elimination of either the elements of water or those of a fatty acid.

A schematic reaction mechanism for the formation of steradienes is shown in the following equation in which R′ is the sterol moiety side chain (Figure 2.18). Sterenes are formed during refining and bleaching processes, and their formation depends mainly on the amount of added earth and its acidity, and is greatly influenced by temperature (Zschau, 2001) The presence of sterenes in an olive oil is certain proof that the oil has been admixed with a refined oil. However, sterenes may be removed from specially prepared refined oils by appropriate procedures. Neutralization, bleaching and deodorization cause a slight reduction in policyclic triterpenes and their artefacts and the magnitude of the loss depends on the conditions used.

The analysis of the sterenes is complementary to that of the *trans* fatty acids and vice versa in the detection of refined oil.

2.13.3 Erythrodiol and uvaol

These two pentacyclic triterpenes are found concentrated in the skin of the olive fruit. Extraction processes employing only physical means results in an olive oil with a low concentration of the two triterpenic diols, their highest limit having been fixed at 4.5% of the overall sterol concentration for oils 1–6. Olive-residue oils obtained by organic solvent extraction of pomace contain large amounts of uvaol and erythrodiol, their official limits respectively being >4.5 and >12% for olive-residue and olive-residue refined oil (see Table 2.2).

Beside olive-residue oil the two triterpene diols can be found in grapeseed oil, a finding that could be an element to account for in the event of suspected sophistication.

2.14 Chlorinated solvents and aromatic hydrocarbons

Olive oil can be contaminated by organic solvents and aromatic hydrocarbons during storage or during production or even at the stage of oleosome development in the olives hanging on the tree (Kotiaho *et al.*, 1995; Biedermann *et al.*, 1995).

Aromatic hydrocarbon contamination of olive oil was first reported back in the 1970s. Since then there have been occasional reports of the presence and identification of polycyclic aromatic hydrocarbons in olive oil, which has caused the consumer serious concern (Menichini *et al.*, 1991).

It has been found that: 1. the olives themselves absorb hydrocarbons from the air; 2. the transporting and storage of the olives causes an increased concentration of aromatics, possibly as a result of further uptake from exhaust gases of engines and the air in the oil mills. However, it was not proven that all the aromatics found in the olive oil arose from absorption from the air. In fact the concentration of styrene was observed to increase during the storage of the crushed olives at room temperature, but this was thought to stem from plant metabolism. However, other reports explain the presence of styrene in olive oil as being due to contamination from containers made of polymers containing the styrene monomer; thus, if the aromatics and halogenated solvents found in olive oil are not the result of malevolent intention, contamination cannot be considered a fraud.

A comment is that olive oil produced in a 'naturally' contaminated area must contain contaminants, unless the oil has been subjected to deodorization. Thus an appropriate contaminant content is a sort of purity criterion for olive oil.

The legal limit introduced by regulations although restricted to halogenated solvents ought to be considered an appropriate measure towards protecting consumer health and helping olive oil producers to find the causes of heavy, dangerous contamination.

2.15 Fatty acids at the glycerol 2-position by lipase method

This method permits a clear distinction between a natural olive oil and an olive oil that has undergone esterification or inter-esterification processes, by determining the distribution of fatty acids on the three positions of glycerol.

Lipases catalyse the following reactions:

$$\text{triglyceride} + H_2O \Rightarrow \text{1,2-diglyceride} + \text{fatty acid}$$
$$\text{1,2-diglyceride} + H_2O \Rightarrow \text{2-glyceride} + \text{fatty acid}$$

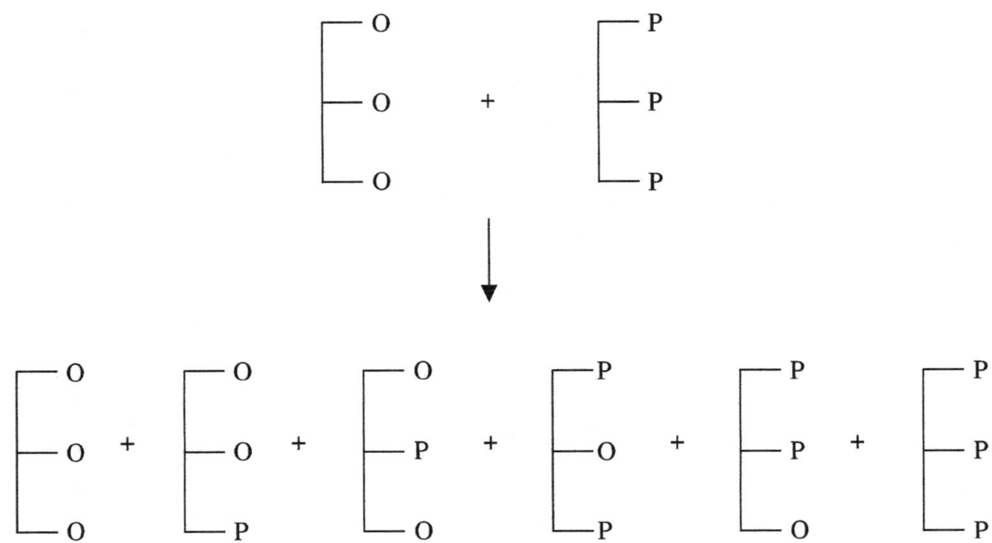

Figure 2.19 Inter-esterification of triglycerides.

Thus, the fatty acyl groups in the terminal positions of triglycerides are preferentially removed with little displacement of those in the 2-position. This positional specificity is independent of the nature of fatty acids at positions 1 and 3 except when the two latter positions are occupied by short chain acids. The formed monoglycerides are separated and transesterified by methanolysis to yield the methyl esters eventually analysed by gas–liquid chromatography. For olive oil, as for most vegetable oils, the saturated fatty acids are almost exclusively concentrated at the 1,3-positions, and almost absent at the 2-position where the oleic and linoleic acids are concentrated. On the contrary, for triglycerides obtained by direct esterification between glycerol and fatty acids or for oil that has undergone inter-esterification, the fatty acid distribution at the glycerol positions are determined only by chance and by the overall fatty acid composition.

As an example, let us consider the fate of triolein (OOO) and tripalmitin (PPP) when the two triglycerides are inter-esterified Figure 2.19 shows the redistribution of the two fatty acids in the six possible triacylglycerols. The official limits of saturated acids at position 2 are determined as $\leq 1.3\%$ for virgin oils, $\leq 1.5\%$ for refined and olive oils and ≤ 1.8–2.0% for olive-residue oils (Table 2.2). The higher limit of 2%, compared with the 1.8 value assigned to crude olive-residue oil for types of oils containing refined olive oil, can be explained as an allowance for triglycerides that can undergo inter-esterification, although limited, by the simple action of heat (Cmolik and Pokorny, 2000).

After inter-esterification an olive oil will present higher concentrations of saturated fatty acids at the 2-position than in the original oil, and this increase can be correlated with the overall percentage of fatty acids. In a real case like that described by Gavriilidou and Boskou (1991) for olive oil–tristearin blends, it is shown how the percentage of saturated fatty acids at the *sn*-2 position of monoglycerides increases significantly in relation to the blend ratios (Table 2.4).

Inter-esterification has many industrial applications such as the production of structured lipids and the manipulation of the physical properties of oils. This method is therefore intended to check the overall genuine authenticity of any commercial olive oil.

Table 2.4 Fatty acid composition of olive oil-glycerol tristearate blends and proportion of fatty acids at 2-position after inter-esterification

	Fatty acids (%)			
	16:0	18:0	18:1 + 18:2	Others
75:25 olive oil–tristearin triglycerides	18	16.3	64.4	1.4
2-monoglycerides before inter-esterification	10.6	17.6	72.5	0.2
2-monoglycerides after inter-esterification	19.1	18.4	61.7	0.8
80:20 olive oil–tristearin triglycerides	15.2	14.7	68.9	1.2
2-monoglycerides before inter-esterification	9.1	13.7	76.5	0.8
2-monoglycerides after inter-esterification	16.1	16.5	66.5	0.9

2.16 Waxes and olive-residue oil

Long chain aliphatic esters, commonly called waxes, are components of epicuticular waxes of the olive fruits (Bianchi and Vlahov, 1994). Whilst in virgin olive oils the wax concentration is negligible, in olive-residue oil the wax content is considerable. Thus, the presence of long chain esters in olive oil is evidence of the presence of solvent extracted olive oil, also commonly called 'sansa' olive oil. The detection and quantification of waxes in olive oil is an official method. As shown in Table 2.2, virgin olive oils fit for consumption must contain less than 250 mg/kg of waxes; the limits are 350 mg/kg for categories 4,5 and 6, whilst for olive-residue oil the amount is expected to be over 350 mg/kg.

Studies on the chemical structure of olive wax esters have shown that the homologues present in olive-pomace oil are almost entirely esters of oleic acid with long chain alkanol constituting the homologous series C40, C42, C44, C46. Odd-chain esters identified in the oil were esters of oleic acid with C23, C25, C27 alcohols. Gas chromatography and mass spectrometry analysis has shown that each carbon chain of the esters in made up of a single isomer in which the acyl moiety is that of oleic acid (Bianchi *et al.*, 1994). Thus, for example, ester C44 was found to be made up of the couple acid-alcohol C18:1 and C26, whereas other possible isomers such as C16:1–C28 were not detected. This is unusual if it is compared with the composition of epicuticular ester fractions of oil seeds for which, in cases studied, each ester chain was composed of several positional isomers of the ester group.

This finding may represent a useful element for the further improvement of this analytical method, which would possibly permit the detection of olive oil sophistication with seed oils.

2.17 Panel test for organoleptic analysis

This method has the main aim of detecting attributes and defects, and measuring their intensity, for the classification of the various categories of virgin olive oils (Angerosa, 2001). The sensory attributes perceived by the consumer arise from the stimulation of gustatory and olfactory receptors from a large number of volatile and some non-volatile compounds such as simple and combined phenols. The intensity of each sensation is related to the concentration of chemical compounds identified in the volatile fraction of the oil.

The volatile fraction of good quality oils is mainly formed by compounds produced enzymatically from polyunsaturated fatty acids through the lipoxygenase (LOX) pathway. Aldehydes (C6), alcohols (C6) and their corresponding esters are the most abundant products (Figure 2.7).

The unpleasant odours of virgin olive oils of poorer categories derive from saturated and monounsaturated aldehydes (C5–C9), some dienals, C5

VIRGIN OLIVE OIL

PROFILE SHEET
OLFACTORY-GUSTATORY-TACTILE NOTES

	0	1	2	3	4	5
Olive fruity (ripe and green)						
Apple						
Other ripe fruit						
Green (leaves, grass)						
Bitter						
Pungent						
Sweet						
Other allowable attribute(s) (Specify.....................)						
Sour/Winey/Vinegary/Acid						
Rough						
Metallic						
Mustiness/humidity						
Muddy sediment						
Fusty ("Atrojado")						
Rancid						
Other unallowable attribute(s) (Specify......................)						

1 Barely perceptible
2 Slight perceptible
3 Average
5 Extreme

GRADING TABLE

DEFECTS	CHARACTERISTICS	OVERALL MARK: POINTS
None	Olive fruity Olive fruity and fruitness of other fresh fruit	9 8 7
Barely perceptible	Weak fruitness of any type	6
Slight perceptible	Rather imperfect fruitness, anomalous odours and tastes	5
Considerable, on the border of acceptability	Clearly imperfect, unpleasant odours and tastes	4
Great and/or serious, clearly perceptible	Totally inadmissible odours and tastes for consumption	3 2 1

Figure 2.20 Profile score sheet for olive oils.

branched aldehydes and some C8 ketones, all C6 compounds present at low concentration in good quality oils. The odour of C6 compounds, the most abundant being *trans*-2-hexenal, is reminiscent of leaves, unripe fruit and just-cut grass, and has been related to 'green' perceptions of virgin olive oil of good quality. Also *cis*-3-hexen-1-ol, hexyl acetate and *cis*-3-hexenyl acetate contribute to emphasising 'green' perceptions. Hexanal gives a 'sweet green' sensation and plays an essential role in the formation of several attributes. Bitter and pungent sensory notes have been correlated with phenolic substances and 1-penten-3-one.

The method uses a group of 8 to 12 persons suitably trained to identify, and to measure, the intensity of the different positive and negative sensations perceived. The tasters rate the intensity of each attribute on a profile score sheet with a scale of 0 to 5 and the overall grading for the characteristics of the oil on a 9-point scale (9 for exceptional characteristics, 1 for the worst (Figure 2.20, Angerosa 2001)).

The initial EC regulation (EC Council, 1991) establishes 6.5 as the minimum score to classify an oil as extra virgin. The other categories are, from a sensory point of view, identified by smaller figures where the values depend on the degree of defective odour perceived with high intensity, such as rancid, musty, fusty, muddy sediment, winey attributes. Table 2.5 summarizes the scores for the different categories.

The poor reproducibility of the overall grading scores complained of in many countries, and mainly due to ineffective training, induced the International Olive Oil Council (IOOC) to revise the method for the organoleptic evaluation of virgin olive oil, and develop a new methodology that considers mainly negative attributes (e.g. fusty, musty, muddy sediment, winey–vinegary, metallic and rancid) that can usually be detected in virgin olive oils and, on a positive note, only fruity, bitter and pungent sensations (Figure 2.21).

Tasters are requested to rate the intensity of each attribute on an unstructured scale 10 cm long. Statistical procedures are applied to the intensity data, expressed as centimetres, to calculate the median of each negative and positive attribute. Thus olive oil is now classified on the basis of the median of defect

Table 2.5 Scores (overall grading) for each category of virgin olive oil

Scores	Categories
6.5	Extra virgin olive oil
$<6.5 \geq 5.5$	Virgin olive oil
$<5.5 \geq 3.5$	Ordinary virgin olive oil
<3.5	Lampante virgin olive oil

PROFILE SHEET

PERCEPTION OF DEFECTS: **INTENSITY**

Fusty

Musty

Winey - Vinegary - Acid - Sour

Muddy sediment

Metallic

Rancid

Others (specify)

PERCEPTION OF POSITIVE ATTRIBUTES:

Fruity

Bitter

Pungent

Figure 2.21 Evaluation sheet adopted by the International Olive Oil Council for the revised panel test grading.

perceived in terms of strongest strength and of fruity attributes for extra virgin and virgin types.

Oils showing a median of the bitter and/or pungent attribute of more than 5.0 should be addressed to blending. The calculation of the robust coefficient of variation provides a measure of the reliability of panel tasters according to data reported in Table 2.6.

Table 2.6 Median values of defects and fruity aroma, and their corresponding robust coefficients of variation, in relation to virgin olive oil categories

Median of defects	Robust coefficient of variation %	Median of fruity aroma	Robust coefficient of variation %	Olive oil category
0	≤20	>0	≤10	Extra virgin
>0 ≤ 2.5	≤20	>0	≤10	Virgin
>2.5 ≤ 6.0	≤20	0		Ordinary
>6.0	≤10	0		Lampante

Acknowledgements

The section 'Panel test for organoleptic analysis' was adapted from a contribution of F. Angerosa. L. Giansante is thanked for help with the literature search. The review of the first draft was made by M.R. Mucciarella and L. Di Giacinto.

References

Angerosa, F. (2001) Sensory quality of olive oils, in *Handbook of Olive Oil* (eds J. Harwood and R. Aparicio), Aspen Publishers, Inc., Gaithersburg, Maryland, USA, pp. 355–392.

Bianchi, G. (1999) Extraction systems and olive oil. *OCL*, **6**, 49–55.

Bianchi, G. and Vlahov, G. (1994) Composition of lipid classes in the morphologically different parts of the olive fruit, *cv*. Coratina (*Olea europaea* Linn.). *Fat Sci. Technol.*, **96**, 72–77.

Bianchi, G., Tava, A., Vlahov, G. and Pozzi, N. (1994) Chemical structure of long-chain esters form 'sansa' olive oil. *J. Am. Oil Chem. Soc.*, **71**, 365–369.

Biedermann, M., Grob, K. and Morchio, G. (1995) On the origin of benzene, toluene, ethylbenzene and xylene in extra virgin olive oil. *Z. Lebensm. Unters Forsch.*, **200**, 266–272.

Cmolik, J. and Pokorny, J. (2000) Physical refining of edible oils. *Eur. J. Lipid Sci. Technol.*, **102**, 472–486.

Cucurachi, A. (1989) Parametri della qualità: olio di oliva. *Ital. Agric.*, **126**, 197–204.

Di Giovacchino, L. (2000) Technological aspects, in *Handbook of Olive Oil* (eds J. Harwood and R. Aparicio), Aspen Publishers, Inc., Gaithersburg, Maryland, USA, pp. 17–59.

EC Council (1991) *Characteristics of Olive Oil and Olive-residue Oil and on the Relevant Methods of Analysis*, Regulation 2568/91. Official Journal 248, 5 September 1991.

Firestone, D. (1994) Liquid chromatographic method for determination of triglycerides in vegetable oils in terms of their partition numbers: summary of collaborative study. *J. AOAC Int.*, **77**, 954–957.

Firestone, D. and Sheppard, A. (1992) Determination of *trans* fatty acids, in *Advances in Lipid Methodology One* (ed. W.W. Christie), The Oily Press Ltd, Ayr, Scotland, pp. 273–322.

Flor, R.V., Hecking, L.T. and Martin, B.D. (1993) Development of high-performance liquid chromatography criteria for determination of grades of commercial olive oils. Part I. The normal ranges for the triacylglycerols. *J. Am. Oil Chem. Soc.*, **70**, 199–203.

Frankel, E.N. (1991) Recent advances in lipid oxidation. *J. Sci. Food Agric.*, **54**, 495–511.

Gavriilidou, V. and Boskou, D. (1991) Chemical interesterification of olive oil tristearin blends for margarines. *Int. J. Food Sci. Technol.*, **26**, 451–456.

Hamilton, R.J., Kalu, C., Prisk, E., Padley, F.B. and Pierce, H. (1997) Chemistry of free radicals in lipids. *Food Chem.*, **60**, 193–199.

Hatanaka, A. (1996) The fresh green odor emitted by plants. *Food Rev. Int.*, **12**, 303–350.

Hayakawa, K.-I. (1967) A method for calculating the ratio of each possible type of triglyceride in natural fat. *J. Am. Oil Chem. Soc.*, **44**, 354–356.

Hoffmann, G. (1970) Keeping properties of oils: are they analytically predictable? *Chem. Ind., London*, pp. 729–732.

Kiritsakis, A. and Christie, W.W. (2000) Analysis of edible oils, in *Handbook of Olive Oil* (eds J. Harwood and R. Aparicio), Aspen Publishers, Inc., Gaithersburg, Maryland, USA, pp. 129–158.

Kotiaho, T., Gylling, S., Lunding, A. and Lauritsen, F.R. (1995) Direct determination of styrene and tetrachloroethylene in olive oil by membrane inlet mass spectrometry. *J. Agric. Food Chem.*, **43**, 928–930.

Leon-Camacho, M., Ruiz-Mendez, M.V., Graciani-Constante, M., Graciani-Costante, E. (2001) Kinetics of *cis-trans* isomerization of linoleic acid in the deodorization and/or physical refining of edible fats. *Eur. J. Lipid Sci. Technol.*, **103**, 85–92.

Menichini, E., Bocca, A., Merli, F., Ianni, D. and Monfredini, F. (1991) Polycyclic aromatic hydrocarbons in olive oils on the Italian market. *Food Addit. Contam.*, **8**, 363–369.

Olias, J.M., Perez, A.G., Rios, J.J. and Sanz, L.C. (1993) Aroma of virgin olive oil: biogenesis of the 'green' odor notes. *J. Agric. Food Chem.*, **41**, 2368–2373.

Porter, N.A., Caldwell, S.E. and Mills, K.A. (1995) Mechanisms of free radical oxidation of unsaturated lipids. *Lipids*, **30**, 277–290.

Rossell, J.B. (1989) Measurement of rancidity, in *Foods*, 2nd Ed (eds J.C. Allen and R.J. Hamilton), Elsevier Applied Science, London, pp. 23–52.

Salas, J.J., Williams, M., Harwood, J.L. and Sanchez, J. (1999) Lipoxygenase activity in olive (*Olea europaea*) fruit. *J. Am. Oil Chem. Soc.*, **76**, 1163–1168.

Smith, M.B. and March, J. (2001) *Advanced Organic Chemistry*, 5th edn. J. Wiley and Sons, New York, Chichester, Weinheim, Brisbane, Singapore, Toronto, pp. 770–773.

Tulloch, A.P. and Craig, B.M. (1964) Determination of double bond position in unsaturated triglycerides by analysis of the oxidation products by gas liquid chromatography. *J. Am. Oil Chem. Soc.*, **41**, 322–326.

Vander Wal, R.J. (1960) Calculation of the distribution of the saturated and unsaturated acyl groups in fats, from pancreatic lipase hydrolysis data. *J. Am. Oil Chem. Soc.*, **37**, 18–20.

Vander Wal, R.J. (1963) The determination of glyceride structure. *J. Am. Oil Chem. Soc.*, **40**, 242–247.

Zschau, W. (2001) Bleaching of edible fats and oils IX. Legal and analytical aspects of bleaching. From the working group 'Technologies of industrial extraction and processing of edible fats'. *Eur. J. Lipid Sci. Technol.*, **103**, 117–122.

3 Authentication of cocoa butter

Colin Crews

3.1 Introduction

Cocoa butter is derived from the tree *Theobroma cacao*, which grows in several tropical areas, including Indonesia, the Ivory Coast, Malaysia, New Guinea and Brazil, which dominate the trade. The seeds of the tree, known as cocoa beans, were first consumed in the form of a drink prepared by the Maya and Aztec Indians. Cocoa beans were carried to Europe during the 16th century and the product was developed into the sweetened solid bar we are familiar with as chocolate. Cocoa butter is used mainly in the manufacture of chocolate confectionery, but it is also popular for applications in cosmetics and as an ingredient of pharmaceutical creams.

The Ivory Coast currently accounts for most (about 40%) of the world's cocoa beans, with about 10% each from Ghana and Indonesia and smaller quantities from Nigeria, Malaysia and Brazil. The production industry is developing rapidly in Indonesia and Brazil, where the crop is grown on plantations under a relatively modern regimented system. The introduction of processing plants in these countries has lead to a decline in the quantity of beans exported from them and a corresponding rise in the export of cocoa powder and cocoa butter. In contrast, much African production remains labour intensive. The market share of the various cocoa growing regions is rather unstable, being affected in several areas by drought, disease and political instability.

The producing countries have had long trade relationships with the Western European nations that formerly colonized them and with the large market in the USA. The Netherlands and the USA are the largest importers of cocoa butter.

Cocoa butter, which forms about 45% of the bean, is extracted by removing the beans from their pods and allowing them to ferment before they are dried, roasted, shelled and ground to a paste known as 'cocoa liquor' or 'cocoa mass'. These stages are complex and most are performed by skilled local labour operating in the open air, where the process must be attuned to effects of the changing weather, which can alter the composition and also allow microbiological infection to take hold. The fermented beans are normally exported for processing, but the producing countries are beginning to develop their own plants. The cocoa liquor is extracted from the kernel or 'nibs' by hydraulic pressing, screw expelling or solvent extraction to produce cocoa butter. The solid residue remaining from pressing, which still contains a quantity of fat, is 'cocoa powder'.

Pressing of liquor made from good quality nibs gives the best quality cocoa butter, which is designated 'pure prime pressed', but butter produced by expeller pressing of good quality nibs is almost equivalent (Timms and Stewart, 1999). Expeller pressing is often applied to poorer grade nibs and may be assisted by steam treatment. It is used primarily to obtain cocoa butter as opposed to both cocoa butter and cocoa powder. Solvent-extracted cocoa butter is of lower quality as it is produced from poorer quality raw materials such as the residues from pressing, and because the solvent extracts some undesirable components from the nibs. This butter is usually inferior in colour with an unpleasant taste and requires degumming and deodorization. Deodorization is usually achieved by passing steam though the butter, under vacuum.

The flavour of cocoa butter is determined by both the geographical origin of the beans and the deodorization conditions. Deodorization reduces the levels of free fatty acids but also some antioxidant compounds such as tocopherols. Deodorized butters are therefore often blended with expressed cocoa butter for better stability of the product.

The physical composition of cocoa butter is defined under different trade standards in different countries, which typically specify a range of acceptable parameters, such as colour, odour and flavour, a range of physically determined values, such as the content of free fatty acid, unsaponifiable matter, peroxides, cooling behaviour and moisture. Important definitions used to ensure that the product is pure and unadulterated are provided by the *Codex Alimentarius* (*Codex*) which states that cocoa butter is 'the fat produced from one or more of cocoa beans, cocoa nib, cocoa mass, cocoa presscake, expeller presscake or cocoa dust by a mechanical process and/or with the aid of permissible solvents. Cocoa butter shall not contain shell fat or germ fat in excess of the proportion in which they occur in the whole bean.' Within the *Codex* standard cocoa butter may be subjected to some forms of processing. It may be filtered, centrifuged, degummed, deodorized, neutralized and bleached (*Codex Alimentarius* Commission, 1981).

The *Codex* standard specifies the following categories of cocoa butter:

1. *Press cocoa butter* is the fat that is obtained by pressure from cocoa nib or cocoa mass (liquor) and treated only by filtering, centrifugation, degumming and deodorization by normal methods.
2. *Expeller cocoa butter* is the fat prepared by the expeller process from cocoa beans singly or in combination with cocoa nib, cocoa mass, cocoa presscake and low fat cocoa presscake. It may be treated only by filtering, centrifugation, degumming and deodorization by normal methods.
3. *Solvent extracted cocoa butter* is the fat obtained by extraction with permitted solvents from cocoa beans and/or from the other raw materials. It shall have been treated by filtering, centrifugation, degumming and deodorization by normal methods.

4. *Refined cocoa butter* is the fat obtained by any of the means described above that has been treated as indicated above and also by neutralizing and/or bleaching.

Cocoa butters have a natural variation in physical properties related to the triacylglycerol structure: Malaysian, Indian and Indonesian butters are harder than those from Africa, and Brazilian butters are the softest. The hardness of typical butters from some continents has changed over the years (Timms and Stewart, 1999). Because the hardness affects the processing required for chocolate manufacture, suppliers of cocoa butter to that trade blend butters to attempt to produce a uniform product.

3.2 Authenticity issues

Because of the comparatively high price of cocoa butter, there is economic gain to be had from the adulteration of what is a premium speciality fat. Other incentives arising from, for example, difficulties in selling inferior grades of fat, and the attraction of recovering fat from 'waste' materials, such as shell and dust, may encourage fraud.

There are two aspects to cocoa butter authenticity. First, the butter as sold to the trade for chocolate production may be of an inferior quality and/or identity to that claimed by the producer. In this situation the consumer will also be misled as a matter of consequence. Second, part of the cocoa butter component of chocolate may be replaced by non-cocoa fats in contravention of legislation or labelling. Whilst this is in the strictest terms an adulteration of the chocolate product the issue has long been discussed in terms of cocoa butter authenticity and will be given due attention in this chapter.

3.2.1 Cocoa butter quality

There has been some history of cocoa butter adulteration. For example, in 1875 a French writer described the substitution of olive oil, sweet almond oil, egg yolks or suet of veal or mutton for the cocoa butter in chocolate (Riant, 1875 in Coe and Coe, 1996). Today the quality of cocoa butter can be affected by the fraudulent addition of foreign fats to cocoa butter or from the misrepresentation of poorer grades of butter as premium product. Thus cocoa butter might be adulterated with hardened vegetable oils, or more likely fat extracted from the cocoa bean shell. Cocoa butter supplied as a premium grade such as 'pure prime pressed' may contain a proportion of added inferior grade butter such as that produced by solvent extraction, or butter from a different country to that specified, or butters with a high diacylglycerol content, or fractionated cocoa butter.

Alternative vegetable fat formulations intended legitimately to replace a small proportion of the cocoa butter in chocolate are difficult to distinguish from cocoa butter without careful study. However, they have not found favour as direct adulterants of cocoa butter as supplied to chocolate manufacturers on account of the fact that they are themselves speciality fats of considerable cost. The major components of non-cocoa fats used in chocolate manufacture are blends made from fats derived from tropical trees which are equally or more difficult to cultivate and harvest than is cocoa butter. As a result of this they are for the most part equally expensive. The appeal of the alternative fats to the confectioner is that they enable a reduction in the cost of the manufacture of chocolate. Therefore it is at this stage that their use as 'adulterants' becomes relevant.

Consumer demands for more specialized chocolate products have led to some newer authenticity issues. There is a growing demand for organically farmed food products, which might be difficult to monitor in a food where the ingredients are grown at considerable distance from the site of product manufacture. Also, *Theobroma cacao* is prone to insect attack and disease and the benefits from the use of chemical agents are high.

Many consumers also wish to be made aware of any genetic modification to the crop. Genetic modification might be beneficial for cocoa butter production as the characteristics of the butter could be modified in the growing bean, and also resistance to pests and diseases might be introduced to the plant. Analytical methods are required to detect such modification, in both the raw materials and the processed product.

3.2.2 Geographical origin

The variations in physical properties, particularly hardness, in beans from different continents means that the cocoa butter processor and chocolate manufacturer must be sure of the origin of the bean. The flavour of cocoa butter is also determined by the geographical origin of the beans and the climate in which they are grown. There are increasing consumer demands for speciality foods, such as wines, olive oil, coffee and cocoa butter produced from named geographical areas or plant varieties, and while most manufacturers usually blend different cocoas before roasting (Dand, 1993) a few chocolate manufacturers have offered premium products made from cocoa butter of stated geographical origin. Also the quality of the chocolate product in terms of, for example, flavour and bloom resistance will depend on the country of origin, and so the chocolate manufacturer has an interest in the authenticity of statements regarding the origin of the cocoa butters used in confectionery.

3.3 Cocoa butter alternatives

The composition of chocolate is subject to a number of international laws that have relevance to the composition and authenticity of the cocoa butter used in its manufacture. The accession of the UK, Denmark and Ireland to the European Community (EC) prompted the introduction in 1973 of an EC directive on the composition, definition and labelling of chocolate that did not permit the addition of non-cocoa fats to chocolate. However, substitution of part of the cocoa butter by non-cocoa vegetable fats was allowed in seven countries, including the UK, but was banned in eight countries, including Belgium, France, Germany and The Netherlands.

The situation was reviewed frequently until the introduction of a new EC directive in 2000 for implementation by August 2003 (European Community, 2000), which authorizes the addition of specific non-cocoa vegetable fats at up to 5% of the total weight of the finished product. The intention of this directive was to ease the free movement of goods, provide a definition of chocolate for sale, define packaging and labelling requirements and inform the consumer of the products' contents in term of the vegetable origin of the fats used.

The effects of the legislation are far-reaching and its introduction complicated by the effect on the demand for cocoa beans and alternative fats. The consequent economic impact on the countries producing cocoa butter has been the subject of much discussion (European Fair Trade Association, 1997). The EC has to meet obligations under international agreements to promote the consumption of cocoa products in their countries and to deal with the problem of country-specific products, such as milk chocolate in the UK. Technical difficulties in measuring the level of addition of non-cocoa vegetables fats to chocolate, necessary to monitor and enforce compliance with the directive, have promoted a great deal of scientific research.

Non-cocoa fats are added to certain chocolates for a number of reasons. Their introduction was prompted by a sharp rise in the cost of cocoa butter in the 1960s which coincided with the emergence of technologies suitable to analyse butter composition and produce substitute fats. Principally, chocolate manufacture can be made more economical by using more stable processing conditions when other fats are added. The variations in processing required by changes in the chemical composition and physical properties of different batches of cocoa butters, and the effects of erratic harvests, can be ameliorated by the incorporation of the tailored non-cocoa fats.

The non-cocoa fats used in confectionery are mixtures known as cocoa butter alternatives (CBAs), of which the most important are cocoa butter equivalents (CBEs). These are formulated from non-hydrogenated fat fractions with a triacylglycerol composition almost identical with cocoa butter and which are miscible with cocoa butter in all proportions. Other alternative fats such as cocoa butter replacers (CBRs) and cocoa butter substitutes (CBSs) are used, particularly in the manufacture of specialized forms of chocolate application such as coatings.

Triacylglycerols are the most important components of both cocoa butter and CBAs. In cocoa butter they are responsible for the characteristic rapid melting behaviour which adds to the appeal of chocolate. Cocoa butter triacylglycerols are simple, containing almost exclusively palmitic (P), stearic (S) and oleic (O) acids, with a lower proportion of linoleic acid (L) in the symmetrical arrangement 1,3-dipalmitoyl-2-oleoyl-*sn*-glycerol (POP), 1-palmitoyl-2-oleoyl-3-stearoyl-*sn*-glycerol (POS) and 1,3-distearoyl-2-oleoyl-*sn*-glycerol (SOS) along the skeletal carbons. This pattern is mimicked in CBEs by blending non-cocoa fats in particular ratios.

Certain tropical fats are identified in the 2000 EC directive for use in the manufacture of chocolate. They must by definition be fats low in lauric acid (non-lauric), rich in POP, POS and SOS triacylglycerols, miscible with cocoa butter, and obtained only by refining and fractionation. Trans-esterification and enzyme modification are excluded. Six fats are specified: illipe (known also as Borneo tallow or Tengkawang) which is derived from *Shorea* species; palm-oil from *Elaeis guineenis* or *E. olifera*; sal from *Shorea robusta*; shea from *Butyrospermum parkii*; kokum gurgi from *Garcinia indica*; and mango kernel from *Mangifera indica*. As certain fractions of these fats closely resemble cocoa butter in all respects there is some potential for them to be added to it fraudulently without detection.

The fats which are used to produce CBAs are refined to reduce free acids and gums, etc. by degumming, bleaching and deodorizing. A typical CBA will be produced by mixing illipe butter with shea and palm kernel. Illipe is similar to cocoa butter but harder, but shea and palm kernel oils are too soft and are therefore subjected to fractional crystallization from solvent. This produces triacylglycerol mixtures of specific composition known as stearines which are harder than the parent fat (Chaudhuri *et al.*, 1983). Various fractions of differing degree of saturation and thus hardness are isolated, particularly from palm, which are known as palm mid-fractions or PMFs (Traitler and Dieffenbacher, 1985). Extensive details of the physical and chemical composition of a number of CBA fats and their applications have been assembled by Wong Soon (1991).

For CBE manufacture, fractionation is preferred to hardening by hydrogenation, which causes melting behaviour unsuitable for bar chocolate. Hydrogenation is however used to produce CBR and CBS from palm or palm kernel oil and rapeseed oil. Fats containing a substantial proportion of lauric acid (lauric fats) which are not compatible with cocoa butter are used in CBS and CBR.

An alternative technique for the production of triacylglycerols is inter-esterification. Here triacylglycerols are modified by altering the arrangement of fatty acids along the glycerol backbone. In most natural fats the 2-position is occupied preferentially by unsaturated acids, thus in cocoa butter the oleic acid is found in this position. Using chemical means the acids may be rearranged in a random distribution. However, the distribution may be directed by using specific lipases (Macrae, 1983). Inter-esterification reactions have often been proposed as a means of producing CBE fats from cheaper fats such as palm oil.

Some degree of enforcement of the rule permitting the addition of up to 5% non-cocoa vegetable fats in chocolate may be based on inspection of factories, recipes and supply records. However, to obtain conclusive evidence of fraudulent addition of non-cocoa fats, a reliable method is required to quantify added fats in chocolate.

Little information has been published on the climatic and geographical factors affecting the composition of CBA fats. It can be assumed, however, that components which vary in cocoa butter with location, etc., also change in other confectionery fats, although these effects are nullified somewhat by refining and fractionation. Comprehensive details of the acylglycerol and fatty acid composition of illipe butters from several *Shorea* species are presented with description of cultivation and harvesting in Blicher-Mathiesen (1994) and some details of the cultivation and uses of shea have been described by Ruiz Méndez and Huesa Lope (1991).

Relatively little information has been published regarding the authenticity of CBA fats. The raw materials of chocolate production are somewhat outside the knowledge of the consuming public but issues of quality are clearly of relevance to the CBA producing industry, and must be addressed within that industry. Given that the cost of most of the tree-borne nuts other than palm often exceeds that of cocoa butter the potential for fraud exists. For example, the shell and fines of illipe and shea, and possibly the other exotic nuts, contain a substantial amount of oil that could be extracted with solvent and used to adulterate the finer grades. Shell oils from illipe and shea have a poorer (higher) free fatty acid content than the kernel and a lower proportion of unsaturated acids (Kershaw, 1987; Kershaw and Hardwick, 1981) and thus would degrade any kernel oil to which they were added.

3.4 Composition and analysis for authenticity

Analytical methods for determining the authenticity of vegetable fats have classically been based on comparison of the composition of some major and minor components. For the most important traded oils, tables of the composition (i.e. profiles) of the major fatty acids and sterols have been of importance and have in some cases been supported by data describing other components such as triacylglycerols, sterol esters, volatiles, waxes and fatty alcohols.

The composition and properties of cocoa butter have been summarized (Chalseri and Dimick, 1987) and extensive details of the composition of cocoa butter and CBA fats have been compiled (Wong Soon, 1991; Lipp and Anklam, 1998a; Lipp *et al.*, 2001). There are serious problems to be considered when applying sets of historical and some contemporary data to the determination of fat authenticity, particularly where this involves the widely used sterol, triacylglycerol and fatty acid data. Improvements in the resolution of chromatography have allowed the separation of components previously regarded as single compounds or unseparable mixtures. Variations in response between different detectors has not often been taken into account when making quantitative measurements. Perhaps of most importance is that a considerable number of published reports are based on the analysis of an unspecified number of samples, occasionally just a single sample, and these are usually of unknown provenance.

The composition of all fats varies with their variety, their geographical origin, the climate and the degree of maturity at harvest. Published data usually describes the composition of the raw fat, extracted with laboratory solvents, whereas product encountered in commerce could be deodorized, refined and fractionated, processes that can have a major effect on all components of the fat. Today there is an increasing awareness of these factors and laboratories responsible for establishing reliable databases will be able to demonstrate the origin of authentic samples and carry out the analysis of statistically significant numbers of samples. It is only through this approach, which is also becoming used more and more in the fields of wine and olive oil authenticity testing, that the analysis of these familiar components will be of value.

All of the classical analytical techniques applied to oils analysis have been applied to studying the authenticity of cocoa butter and considerable attention has been paid to the issue of detecting and quantifying non-cocoa fats in mixtures with cocoa butter and as incorporated into chocolate. A substantial review of these methods has been published (Lipp and Anklam, 1998b).

3.4.1 Acylglycerols

Tables have been published describing ranges of the major triacylglycerols in cocoa butter, CBAs and fats used in the manufacture of CBAs (Chaudhuri *et al.*, 1983; Shukla *et al.*, 1983; Podlaha *et al.*, 1984, 1985; Řezanka and Mareš, 1991; Ruiz Méndez and Huesa Lope, 1991; Wong Soon, 1991; Blicher-Mathiesen, 1994; Shukla, 1995, 1997; Lipp and Anklam, 1998a).

Cocoa butters from Malaya, India and Sri Lanka contain a higher proportion of monounsaturated triacylglycerols than South American butters and are consequently harder. South American butters contain the lowest proportions of monounsaturates and the highest proportions of di- and polyunsaturates. African cocoa butters fall between the South American and Indian types. The triacylglycerol composition is affected by the temperature, sunshine and rainfall during development of the seeds and also by the point in growth at which the seeds are harvested. Postharvest processing, principally roasting and deodorization, have less effect on the triacylglycerol composition but may affect minor components.

Triacylglycerols in cocoa butters have been separated by high performance liquid chromatography (HPLC) or gas chromatography (GC). These methods have been compared by Řezanka and Mareš (1991) who reported some differences in performance from limited data and by Buchgraber *et al.* (2000) who found no significant difference from a large number of samples analysed under conditions carefully selected to unify response factors.

Reverse phase columns are preferred for HPLC with solvent systems based on acetonitrile with dichloromethane, acetone, tetrahydrofuran or methyl tertiary-butyl ether added to reduce polarity. The efficiencies of many stationary phases and solvent systems have been compared (Héron and Tchapla, 1994) and it is considered that polarizable polymeric columns provide the optimum general performance. Ultraviolet (UV) detection has been used with some success (Shukla *et al.*, 1983) but the choice of wavelength affects sensitivity to individual triacylglycerols which in any case absorb poorly in the UV range. Because of this, detection systems based on refractive index detection (Hernandez *et al.*, 1991) and evaporative light scattering detection (ELSD) are more commonly used. ELSD has the significant advantage in permitting a wider range of solvent programming to be used (Robinson and Macrae, 1984; Anklam *et al.*, 1996; Buchgraber *et al.*, 2000) but some light scattering detectors might suffer from lack of linearity and reproducibility with some solvent programmes (Robinson *et al.*, 1985). All detection methods differ in their responses and therefore care should be taken when comparing results from different laboratories.

Enhanced separation of cocoa butter triacylglycerols can be achieved by the use of propionitrile as solvent or by bromination of the triacylglycerols, which also improves UV detector response. By the latter means, Geeraert and de Schepper (1983) were able to distinguish between cocoa butters, PMFs, sal and shea and provide some quantification of CBE mixtures in cocoa butter. Treatment of the stationary phase of thin-layer chromatography (TLC) or HPLC systems with silver ions (argentation) markedly affects the separation of unsaturated compounds, which takes place according to the degree of unsaturation and the position of the double bonds (Christie, 1995; Jeffrey, 1991). Argentation TLC with two-dimensional developments gave rapid and effective separation of CBA triacylglycerols (Jee and Ritchie, 1984) and could be used as a semi-quantitative procedure or an isolation mechanism for cocoa butter and similar fats. The use of argentation HPLC with subsequent capillary GC has been used as a means of quantifying the percentage of triacylglycerol acids in cocoa butter and CBE fats (Neri *et al.*, 1998, 1999). The authors proposed the construction of multiple regression and similar statistical models, which should help to enable the identification and/or the quantification of foreign fats added to cocoa butter, and certainly of CBEs added to chocolate.

Today, mass spectrometry offers an attractive alternative as a detector to HPLC. Newer techniques for linking HPLC systems with mass spectrometers directly via atmospheric pressure chemical ionization (APCI) and electrospray interfaces should see an expansion of this analytical tool in the analysis of confectionery fats, a field in which it has not yet been applied. Triacylglycerols show little fragmentation in APCI, saturated acylglycerols producing only diacylglycerol fragment ions, and unsaturated fats a mixture of these and protonated molecular ions (Byrdwell and Emken, 1995).

GC separations have tended to use comparatively short length (5 to 15 m) low or medium polarity packed or capillary columns operated at high temperatures (about 300°C), with flame ionization detection (FID). Careful consideration must be paid to the injection conditions. Cooled on-column injection techniques are preferred in order to reduce degradation of some triacylglycerols, but there is some evidence that such degradation of some triacylglycerols can occur at high column temperatures. It has also been shown that carrier gas velocity affects the response of the FID detector (Buchgraber *et al.*, 2000).

With low polarity columns, separation is based on the number of carbon atoms, with unsaturation of the acyl substituent not detected. The degree of separation can be improved by the use of medium-polarity columns when the number of double bonds in the acyl molecule does have an effect on the separation. More recently there has been a move towards the use of longer capillary columns but even this does not give a very substantial improvement in separation of triacylglycerols than can be achieved by HPLC.

Triacylglycerol analysis is more important in identifying fat type in confectionery analysis than in many other fields of oil authenticity. Von Klagge and Gupta (1990) found significant differences in the triacylglycerol profiles of single cocoa butter samples from six countries, which were correlated with determinations of the solid fat content. Simoneau *et al.* (1999) were able to distinguish the geographical origin of most of over 50 cocoa butters by plotting POO + PLS versus SOO + SLS. Brazilian butters were shown to contain higher proportions of POO/PLS and SOO/SLS, while Asian butters had lower proportions of POO and SOO. The authors also measured low levels of two commercial CBAs in cocoa butter by plotting POS versus SOS. Triacylglycerol analysis is used to detect and quantify CBEs in chocolate, either in the product or in development work based on laboratory mixtures of cocoa butter and CBAs. Using GC analysis of triacylglycerols, Lipp *et al.* (1996) were able to determine low levels of CBEs in mixtures with cocoa butter. Plots of the ratios of POS versus POP and SOS versus POP have been used to detect low levels of illipe butter in cocoa butter (Biino and Carlesi, 1971).

Current methods for the quantitative determination of vegetable fats in chocolate are based on the measurement of C50, C52 and C54 triacyglycerols, which are present in both cocoa butters and vegetable fats. C54 and C50 triacyglycerols in cocoa butter have a linear relationship. By setting the sum of C50, C52 and C54 triacyglycerols to 100% and plotting the percentage of C50 against the percentage of C54, the level of a non-cocoa fat of known triacylglycerol profile in a mixture with cocoa butter may be calculated (Padley and Timms, 1980). The profiles of triacyglycerols used as commercial CBEs lie within a band on the C54–C50 plane. The boundaries of this band may be determined by the analysis of a small number of CBEs and its range can be used to used to calculate the quantity of an unidentified non-cocoa fat in a mixture with cocoa butter (Young, 1984).

The Padley and Timms (1980) method has been the subject of a collaborative trial which showed the method to be suitable for use by competent laboratories for the analysis of fats that did not contain added milk fat, hazelnut oil or the CBE 'Calvettta', which has a rather different triacylglycerol composition to other CBEs (FSA, 2001).

The precision of the Padley and Timms' (1980) and Young's (1984) procedures has been improved by the incorporation of data derived from sterol degradation product analysis which helps to identify the type of fat present and therefore narrows the range along the C54–C50 band in which the fat falls (Macarthur *et al.*, 2000). These improvements to the method have been proposed for adoption into the *Codex* standard (*Codex Alimentarius* Commission, 2001).

Procedures based on comparison of triacylglycerol compositions are affected by the presence of milk fat in some chocolate products and steps must be taken to account for the contribution of this fat to the triacylglycerol profile (Fincke, 1982; Minim and Cecchi, 1999). Difficulties might also be caused by the addition to chocolate confectionery of other components, such as nuts, biscuit and reprocesseed chocolate. Interference from milk fat and hazelnut oil triacylglycerols can be minimized by measuring the correlations between PPO and SSO and between PSO and PPO in a similar way to the use of correlations of C50, C52 and C54 triacylglycerols. In this way Podlaha *et al.* (1984) studied the composition of cocoa butters prepared in the laboratory from 28 samples of cocoa beans from five geographical areas. The triacylglycerol composition was determined by reverse-phase HPLC using propionitrile as mobile phase. Correlations based on triacylglycerol type, between PPO and SSO, and between PSO and PPO rather than carbon number, were used to detect CBEs in cocoa butters. Plotting the PPO content against the mean number of double bonds in the triacylglycerol enabled some grouping of samples based on geographical location.

Some detail of the sterochemical composition of triacylglycerols can be obtained by measurement of the fatty acid in the 2-position. Acids in this position are usually less saturated than those in the 1- and 3-positions, and are frequently characteristic of particular fats. Measurement of the 2-position acid is achieved by using the enzyme pancreatic lipase to remove the fatty acids in the 1- and 3-positions of the triacylglycerol allowing determination of the remaining acid. Other chemical techniques are able to provide data on the fatty acid present at all three positions based on reaction of triacylglycerol with Grignard reagent (ethyl magnesium bromide) followed by separation of the diacylglycerols by derivatization and HPLC (Ando and Takagi, 1993; Takagi and Ando, 1995; Christie, 1995; Takahashi *et al.*, 2001). To date there has been very little application of this technique in comparing cocoa butter with other fats, but with either the random or specific rearrangement of fatty acids during CBA manufacturing processes, such inter-esterification should make them distinguishable from cocoa butter by sterochemical analysis.

Mono- and diacyglycerols are also present in confectionery fats at low levels, generally 1 to 4%, and are of interest in chocolate manufacture as they have a strong and adverse effect on the crystallization behaviour. Their analysis is of less importance in testing for authenticity. Siew and Ng (1995) have pointed out that the ratios of 1,2-diacyglycerol isomers to 1,3-isomers in palm oil vary very considerably with the maturity of the fruit, its storage and, in particular, the processing conditions used. These variations, particularly if they also occur in cocoa and other butters, will further reduce the value of diacyglycerol analysis in authenticity testing.

Diacyglycerols have been determined by gas chromatography after derivatization to the silyl ethers (Bruschweiler and Dieffenbacher, 1991) but are probably more conveniently separated by HPLC. Liu *et al.* (1993) resolved 1,3-diacylglycerols from the 1,2-positional isomers by normal phase HPLC, but some low molecular weight 1,3-diacylglycerols co-eluted with the 1,2-diacylglycerols of higher molecular weight. Only positional isomers of monoacylglycerols having the same fatty acid could be separated from each other. Using reverse phase HPLC with a polymer chiral column Itabashi *et al.* (1991) has separated dinitrophenylurethanes of diacylglycerols. The derivatives exhibited characteristic fragmentation under online liquid chromatography–mass spectrometry.

3.4.2 Fatty acids

Although vegetable fats contain small quantities of free (non-esterified) fatty acids they are regarded as undesirable components and are largely removed during the early stages of refining. The published fatty acid composition values for vegetable oils and fats therefore invariably includes and refers mainly to the proportion of acids derived from triacylglycerols and sterol esters. Many such tables for cocoa butter, CBAs and CBA component fats have been published (Biino and Carlesi, 1971; Derbesy and Richert, 1979; Fincke, 1980; Baliga and Shitole, 1981; Kershaw, 1987; Chaudhuri *et al.*, 1983; Gaydou and Bouchet, 1984; Gunstone *et al.*, 1994; Thippesawamy and Raina, 1989; Yella Reddy and Prabhakar, 1990, 1994; Harwood, 1991; Ruiz Méndez and Huesa Lope, 1991; Nesareretnam and Razak bin Mohd Ali, 1992; Dencausse *et al.*, 1995; Shukla, 1997; Timms and Stewart, 1999).

The fatty acids esterified as acylglycerols are determined almost exclusively by GC of their methyl esters using recognized techniques (Shantha and Napolitano, 1992; Eder, 1995). A variety of GC columns can be used, the length and polarity of which depend on the detail of composition required, longer and more polar columns being required to separate geometrical isomers of the unsaturated acids. Much attention has been devoted to the means of formation of the methyl esters, which can affect quantitative results. Esterification is normally achieved by acid or base catalysed transmethylation. For the former, solutions of hydrogen chloride, sulphuric acid or boron trifluoride in methanol are most often used, and for base catalysed reactions sodium methoxide is most commonly encountered. Other esterification reagents such as diazomethane and quaternary ammonium salts figure less prominently, particularly in official methods. Alternative esters, such as those of picolinic acid, may be used to aid in the identification of less familiar acids by their characteristic mass spectra.

Fatty acid determination has not often been applied to cocoa butter authenticity in isolation. Wong Soon (1991) showed the addition of illipe to cocoa butter in a model system by measuring the fatty acid composition of mixtures but the change in composition did not reflect the level of addition of illipe. Lipp *et al.* (2001) found differences in the C18:2 content between South American, African and Asian butters. However, determination of fatty acid profiles should be regarded as an important factor to consider, particularly as part of multivariate analytical schemes.

Hardening of fats by hydrogenation can be detected by determining the proportions of *trans*- to *cis*-fatty acids. Some degree of isomerization takes place during refining when dienoic and trienoic mono *trans*-fatty acids are formed, albeit usually at low levels. During hydrogenation monoenoic acids are formed in greater proportions (Ackman *et al.*, 1974). Infrared spectroscopy can be used to quantify *cis*- and *trans*-fatty acid isomers but the sensitivity of this technique is poor. Gas chromatography is the method of choice, enabling the separation of most of the more important isomers. However, attention is required to ensure that optimal separation is achieved of most of the numerous isomers in many fats, linolenic acid alone having eight geometrical isomers. Many separation conditions have been developed. Those specified by the American Oil Chemists' Society and some alternatives have been studied and compared (Duchateau *et al.*, 1996).

3.4.3 Sterols

Although sterols are minor components of fats they have long been considered important in the determination of oil and fat authenticity. The most familiar group are the desmethylsterols, whose structure is based on the cholesterol molecule. Cholesterol makes up a relatively small proportion of vegetable fats, of which the major component of most is β-sitosterol (24-ethylcholesterol). Other sterol groups are the 4-methylsterols and the triterpene alcohols, or 4,4-dimethylsterols. 4-Methyl sterols are of much less importance in determining the authenticity of cocoa butter as they occur at low levels in all of the fats

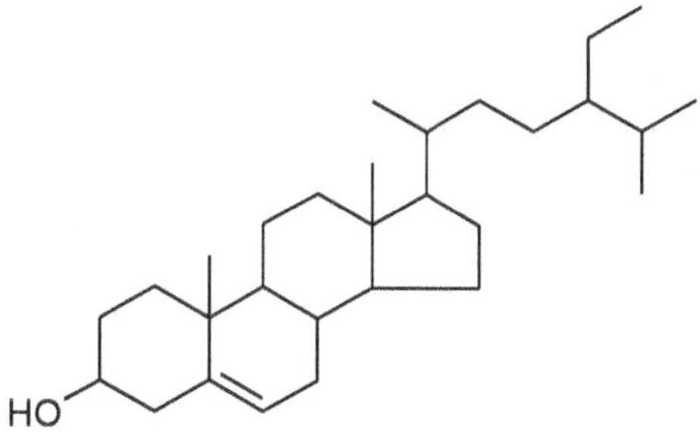

I

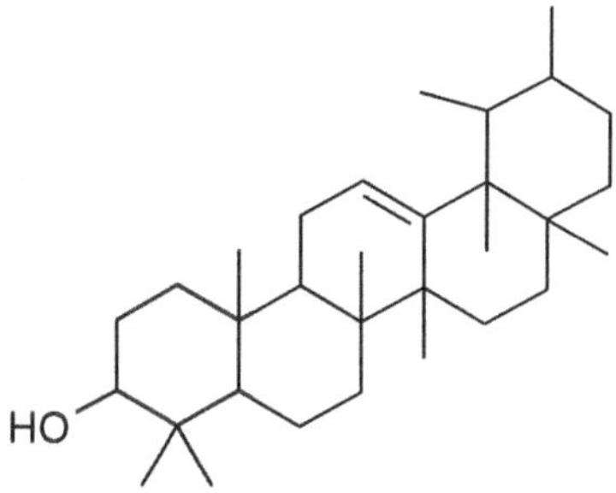

II

Figure 3.1 Two sterols in cocoa butter, β-sitosterol (I) and α-amyrin (II).

of interest and have similar profiles in cocoa butter and CBA fats. The major 4-methylsterols identified in cocoa butter (Itoh *et al.*, 1974) are obtusifoliol, gramisterol and citrostadienol.

The structures of two important sterols, the desmethylsterol β-sitosterol (I) and the triterpene alcohol α-amyrin (II), are shown in Figure 3.1.

Many tables describing the sterol profiles of cocoa butter, CBAs and CBA component fats have been published (Itoh *et al.*, 1974; Derbesy and Richert, 1979; Homberg and Bielefeld, 1982; Staphylakis and Gegiou, 1985a; Homberg and Bielefeld, 1989; Soulier *et al.*, 1990; Frega *et al.*, 1993; Gunstone *et al.*, 1994; Dencausse *et al.*, 1995) but improvements in chromatographic resolution and the identification of GC peaks have made older data of less relevance.

The sterols present in cocoa butter make up about 0.3 to 0.4% of the oil and, as in other oils, they exist as free sterols, esters with fatty acids, and as glucosides and acylated sterol glucosides. Sterol determinations are normally carried out on the saponified material after isolation of the sterol fraction by TLC, although recovery from the TLC plate is lower than can be achieved by the use of HPLC, which also gives improved separation of the desmethylsterols and triterpene alcohols. Gas chromatography can be performed on the non-derivatized sterols provided the column is in good condition but acetylation (which can precede TLC) or silylation gives more consistently reliable resolution and peak shape.

There are difficulties in the lack of an ideal internal standard for sterol determinations. Most used for this purpose are α-cholestanol (dihydrocholestanol), cholestane and betulin. Cholestanol is perhaps the most suitable where its resolution from cholesterol can be demonstrated. Cholestane and betulin, having none and two hydroxyl groups, respectively, do not migrate closely with the sterols on TLC and clearly their derivatization requirements do not match those of the analytes.

Cocoa butter has a high proportion of stigmasterol compared with other confectionery fats. In particular the proportion of stigmasterol in the sterol fraction is particularly low in illipe butter (Biino and Carlesi, 1971). This factor has been used by Bracco *et al.* (1970) and by Homberg and Bielefeld (1982) who reported a significant difference in the ratio of stigmasterol to campesterol when comparing cocoa butter (mean ratio for three samples = 3.18) with ten CBAs and fats used in cocoa butter formulation (range = 0.38 to 1.47). In addition the low stigmasterol content of hazelnut oil and the absence of phytosterols in milk fat means that these common ingredients of some chocolates would not affect the use of stigmasterol:campesterol ratios to detect addition of non-cocoa fats.

Measurement of the proportions of the major cocoa butter sterols have been proposed as a purity check for cocoa butter and for chocolate by Garcia Olmedo and Diaz Marquina (1974) who suggested limit values for the ratios of campesterol, sitosterol and stigmasterol.

It is of some significance that the sterols of shea butter differ considerably from those of cocoa butter. Shea butter is a major component of the most commonly encountered CBEs. It contains a high level of unsaponifiable matter, about 5% compared with less than 1% for cocoa butter. The sterol fraction contains only very low levels of the more familiar desmethysterols (β-sitosterol, campesterol and stigmasterol) but distinctively high proportions (about 40% each) of spinasterol (stigmasta-7,22-dienol) and Δ^7-stigmastenol, and 3–4% of Δ^7-campesterol and stigmastatrienol (Peers, 1977; Artaud *et al.*, 1995). Of equal importance are the high levels of distinctive triterpene alcohols, principally α- and β-amyrins, lupeol and butyrospermol.

Much of the sterol component of shea butter is present as esters of cinnamic acid, which are less readily saponified than esters with glycerol (Peers, 1977); published data for this fat might therefore underestimate the true value. Surprisingly measurement of cinnamic acid has not been used to test for the presence of shea butter in mixtures. Triterpene alcohols similar to those found in shea also occur in sal fat. Homberg and Bielefeld (1982) showed the presence of triterpene alcohols in illipe and sal fats and in commercial CBEs, and their analysis was proposed as a qualitative measure to detect cocoa butter adulteration.

The presence of another triterpene alcohol, taraxasterol, has been reported in shea (Itoh *et al.*, 1979; Dencausse, 1995), and also in illipe and sal fat (Soulier *et al.*, 1990). These relatively minor sterols have been little studied or reported and should be included in the analysis of sterolic fractions of confectionery fats.

Reaction of shea triterpene alcohols with acetic anhydride and sulphuric acid to produce coloured products (Fitelson's reaction) was the basis of a sensitive early test for the presence of shea butter in cocoa butter (Fincke, 1975). Analysis of the triterpene fraction of a commercial cocoa butter by TLC fractionation followed by GC (Fincke, 1976), or argentation TLC followed by GC (Gegiou and Staphylakis, 1985), have been shown to have potential for detecting CBEs in chocolate based on the difference in levels of β-amyrin, butyrospermol and lupeol. Measurement of α-spinasterol and analysis of the triterpene alcohol fraction have been used to indicate the presence of shea butter in confectionery fats and mixtures of shea and cocoa butters at levels of addition as low as 1% (Derbesy and Richert, 1979).

In the analysis of sterols and other components of CBAs, in particular the minor components it must be borne in mind that it is selected fractions of the parent fats obtained by crystallization that are used in confectionery and are potential adulterants. These fat fractions are produced with a view to obtaining desired physical properties, which depend on the triacylglycerol composition, and while they do contain measurable levels of some minor components of varying polarity their composition is likely to vary significantly from that of the parent fat.

3.4.4 Sterol esters

Analysis of the esterified sterol is of value in determining oil and fat authenticity as the sterol composition of the esters is usually more characteristic than that of the total unsaponifiable fraction and also because the sterol esters are less affected by refining. Column chromatography, TLC or HPLC are required to separate free, esterified and glucosidic sterol fractions and, as the low volatility of the intact ester and glucosides makes GC separation difficult, saponification is generally applied to the isolated ester fraction. Staphylakis and Gegiou (1985b), in a comprehensive description of the fractionation of cocoa butter sterol components, characterized cocoa butter as containing 200 mg free sterols, 50 mg steryl esters, 40 mg steryl glucosides and 40 mg acylated steryl glucosides per 100 g unsaponifiable matter and provided details of the sterol composition of each fraction. No methylsterol or triterpene alcohols were found as glycosides.

Kamm *et al.* (2001) analysed intact steryl esters in the presence of silylated sterols by online transfer from HPLC to a 15 m low-polarity high temperature GC column. Steryl esters were quantified against a cholesteryl laurate internal standard and identified by offline thin layer chromatography–gas chromatography–mass spectrometry. When esters of sitosterol, stigmasterol and campesterol with the major cocoa butter acids were measured in cocoa butters from South America, Asia and Africa, no differences were apparent between cocoa butters of different geographical origin, or between deodorized and non-deodorized butters.

No comprehensive comparison has been made of the proportions of the free and esterified sterols in cocoa butter with those of fats likely to be used in cocoa butter adulteration. However Gordon and Griffiths (1992) examined the sterol esters of palm kernel oil by isolation with TLC followed by GC and HPLC. They pointed out the problem of co-elution of triacylglycerols with steryl esters with GC. The characterization of esters of triterpene alcohols in CBA fats might well prove useful where the use of fats containing shea is suspected.

3.4.5 Sterol degradation products

Analysis of the sterols fraction of confectionery fats, particularly when looking for evidence of the addition of non-cocoa fats or refined cocoa butter to the prime product, suffers from the degradation of the sterols during processing. Virtually all of the fats used as CBAs are bleached and deodorized before use. Bleaching is carried out by heating the fat for a few hours in the presence of acidic bleaching earths at a temperature of about 150°C to remove chlorophyll, carotenoids and labile polar materials. Bleaching also destroys tocopherols and a proportion of the desmethylsterols. Triterpene alcohols and Δ-7 sterols are less affected, some undergoing a degree of isomerization. Destruction of sterols by bleaching might be used as a deliberate process to avoid detection of adulteration (Grob *et al.*, 1994).

Sterols are dehydrated to a mixture of disteryl ethers and steroidal hydrocarbons known as steredienes or sterenes, in which the –OH group in the 3-position is replaced by a double bond to give the 3,5-diene. Thus β-sitosterol forms 3,5-stigmastadiene (III) (Figure 3.2). Some isomerization occurs to produce additional isomers such as 2,4- and 4,6-dienes. The rate and extent of formation of these products is dependent on the bleaching conditions and although they can be quantified with relative ease it is not possible to relate these quantities to the amounts of the parent sterol. However, the products are formed in approximately the same proportions as their parent sterols and can give some indication of the identity of the fat.

Detection of stigmastadienes in virgin olive oil at levels in excess of 0.15 mg/kg is regarded under EC regulations as evidence of the presence of refined oils. The test is highly sensitive; stigmastadiene levels in unrefined cocoa butter are well below 0.1 mg/kg whereas in refined butters up to several hundred mg/kg may be present.

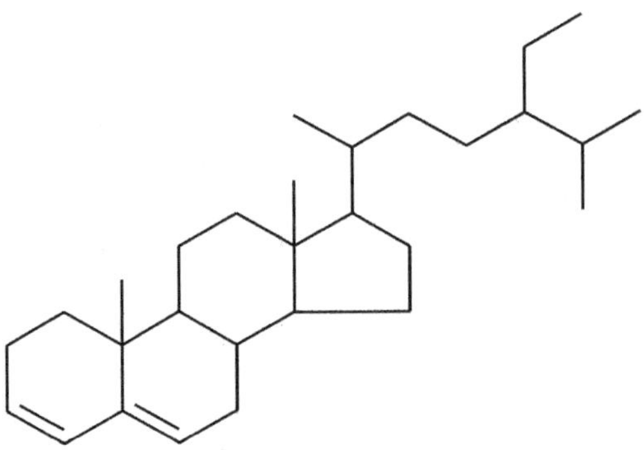

III

Figure 3.2 3,5-stigmastadiene (III).

The quantitative determination of sterenes has been used to detect added CBEs or refining in cocoa butter and to establish the presence of CBAs in retail chocolates (Crews *et al.*, 1997). The high quantity of triterpene alcohols in shea butters means that dehydration products of these sterols are formed in detectable levels when shea is bleached. These triterpene sterenes, which have not be characterized, can be detected in the stearin fraction used in CBEs (Crews *et al.*, 1999).

3.4.6 *Tocopherols*

Tocopherols and tocotrienols are precursors of vitamin E and are important antioxidants in oils. Their reactivity means that they are not stable to many oil processing procedures, including deodorization, which reduces levels by up to 15%. Levels of tocopherols in cocoa butter are usually about 100–300 mg/kg, with the γ-isomer (IV) being the major component (about 90%), but they can be entirely absent (Lipp *et al.*, 2001) (Figure 3.3). Tocotrienols have a similar structure with unsaturation of alternate bonds along the alkyl chain. Only γ-tocotrienol is found in cocoa butter and this at low levels (<5 mg/kg). Palm oil is notably high in tocopherols and tocotrienols, of which α-tocopherol and α-tocotrienol make up 20% to 30% each with most of the remainder as γ-tocotrienol.

Losses of these compounds to light, thermal degradation and mould growth, and their destruction on refining, coupled with difficulties with analytical methodology (Coors, 1991), have limited their application to cocoa butter authenticity. Tocopherols have been determined after saponification of cocoa butter and isolation by TLC followed by derivatization and gas chromatography (Erickson *et al.*, 1973). This early technique did not separate β- and γ-tocopherols but did show small differences in samples from different countries, and that roasting had little effect on the levels or proportions. It was also found that tocopherol levels were much higher in fat extracted from the cocoa bean shell, with α-tocopherol the predominant isomer. As the proportion of shell fat in a sample of cocoa butter pressed from whole beans would be less than 1%, a high tocopherol level in cocoa butter might indicate adulteration with this

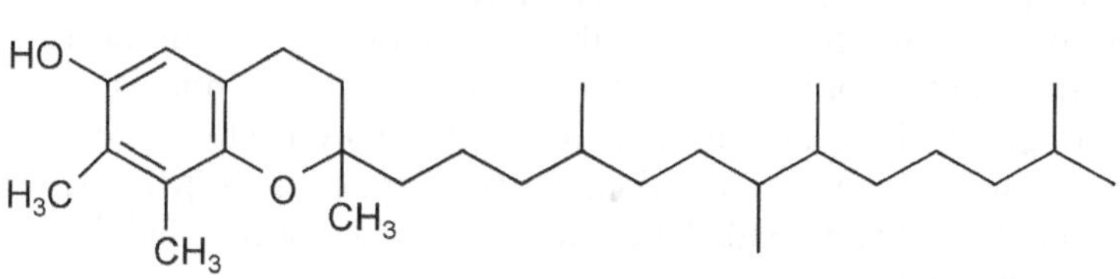

IV

Figure 3.3 γ-Tocopherol (IV).

fat. Today tocopherols and tocotrienols are usually determined by normal phase HPLC with fluorescence detection (Lipp *et al.*, 2001).

3.4.7 Pyrolysis products

Pyrolysis of food samples provides a large number of products, which, after detection by MS and analysis by advanced statistical treatment, can be used to compare different samples. The method is very rapid and does not require chromatographic separation or MS identification of the pyrolysis products. Pyrolysis MS, coupled with multivariate data analysis procedures, has been used to discriminate between cocoa butters of three different continental areas (Radovic *et al.*, 1998). The technique could in some cases separate deodorized from non-deodorized cocoa butters and also show those that have had alkali treatment. The presence of non-cocoa fats did not affect the assay.

3.4.8 Volatile components

As might be expected from a foodstuff of distinct flavour that has undergone roasting during its manufacture, cocoa butter contains a considerable number of volatile components. Carlin *et al.* (1986) identified nearly 100 compounds in roasted and unroasted butters of which pyrazines, thiazoles, oxazoles and pyridines were largely responsible for the flavour. The number and levels of pyrazines can be used to distinguish between roasted and unroasted butters. Volatile components from fats can be extracted in high yield by such techniques as dynamic purging of the headspace vapour into a trap (Pino, 1992) or steam distillation (Pino *et al.*, 1993). These methods require thermal desorption or solvent extraction from a trap, or, in the case of simultaneous distillation–extraction, the steam and solvent vapours are mixed and the solvent collected separately. By such means Pino and Roncal (1992) determined linalool in roasted cocoa butter as a marker of the grade of flavour.

These semi-preparative methods are useful where identification is required but for quantitative and comparative analytical purposes much more rapid sampling techniques, such as automated headspace and solid phase microextraction (SPME), may be preferred. Both of these techniques give similar results for most volatiles. In the former, the vapour above a heated sample is removed by a syringe or gas flushing and injected onto a GC column, either directly or after trapping on a suitable absorbent and thermal desorption. In SPME, the vapour is absorbed on to a suitable bonded medium on a special needle and then injected into the gas chromatogram.

Hernandez and Rutledge (1994a) used dynamic headspace sampling to analyse cocoa mass from 13 geographical locations at different roasting stages. Multivariate statistical methods, using a small number of identified principal components, were successful in differentiating the masses in terms of supplier and roasting step. The approach could also be applied to cocoa butter.

Headspace analysis and SPME methods produce a wealth of chromatographic data and the best approach is to use chemometric analysis of selected chromatographic peaks under which circumstances identification of the individual compounds is not usually necessary. These techniques have been applied with some success to characterize olive oils (Morales and Aparicio, 1993; Morales *et al.*, 1994).

3.4.9 Trace elements

Trace element analysis of foods can be carried out to check for contamination by toxic elements, such as lead and cadmium, or to determine beneficial micronutrients, or as an aid to distinguishing geographical origin. In fats, small numbers of trace elements are measured after digestion of the sample in acid followed atomic absorption spectrophotometry (AAS) or by direct graphite furnace vaporization. An AAS procedure for measuring lead in edible oils and fats has been collaboratively trialed with cocoa butter as a test material (Firestone, 1994).

Kohiyama *et al.* (1992) reported mean levels of nickel, iron and copper of 0.03, 0.30 and 0.04 mg/kg, respectively, in 10 samples of cocoa butter. Baxter *et al.* (2001) found comparable results from the determination of 23 elements in 42 cocoa butters and 22 CBA fats, mostly of known geographical origin and processing history, by the sensitive multi-element technique of inductively coupled MS. No distinction could be made between the cocoa butters based on geographical origin or deodorization, and the very low levels of most elements in the CBA fats meant that their presence in mixtures with cocoa butter could not be detected.

Hernandez *et al.* (1994) used AAS with acid digestion to measure copper, zinc, iron, magnesium and manganese in four South American and 15 African cocoa butters. The much higher levels reported (e.g. iron, 20–50 mg/kg, copper, 11–27 mg/kg) were ascribed to contamination by metals from grinding tools. Principal component analysis could not distinguish between samples from different continents or countries.

3.4.10 Stable isotope ratios

The ratio of certain stable isotopes, principally $^{13}C/^{12}C$ and $^{18}O/^{16}O$, varies with geographical location where climatic effects change the proportions of the isotopes available to plants. Stable isotope measurements can be carried out on the entire sample after pyrolysis or combustion, or on specific compounds after separation by GC. These techniques therefore have potential to identify the geographical origin of cocoa butters and CBA fats although the approach has not yet been applied in earnest.

3.4.11 Physical methods

The quantity of fat from the cocoa shell that ends up in the cocoa butter can be measured by testing with *p*-dimethyl-aminobenzaldehyde, which reacts with dried fruit pulp adhering to the shell. The reaction gives an intense fluorescence and has been used with TLC and fluorescence spectrophotometry (Kleinert, 1964). Shell fat is softer than nib fat because it contains higher levels of linoleic acid (18:2) and its presence softens the butter (Timms and Stewart, 1999). However, its presence can be usually due to poor separation of shell from nib rather than direct adulteration.

A fluorescence under UV illumination of an unidentified compound separated by TLC has been used to detect 5% kokum (possibly unrefined) in mixtures with cocoa butter (Deotale *et al.*, 1990). Identification of this compound and its analysis by more specific techniques might be used to improve current methods of quantifying CBEs in chocolate.

Thermal analysis techniques are used to produce diagrams of the formation of different crystalline forms of cocoa butter induced by heating and cooling, by monitoring changes in enthalpy. The complexity of the triacylglycerol structure of oils is responsible for mixed melting and crystallization profiles, of which the crystallization behaviour provides the most useful data (Md Ali and Dimick, 1994). Thermal analysis is widely used in industry to measure crystalline forms and melting behaviour of cocoa butter as this is of great importance in chocolate manufacture; however, these measurements have had less application in detecting adulteration. The technique can show some differences between butters of different geographical origin and can distinguish between pressed and solvent extracted butters (Bracco *et al.*, 1967; Chevalley *et al.*, 1970). Differential scanning calorimetry, the most widely used form, has been studied as a means of distinguishing between cocoa butter and cocoa butter alternatives (Kerti, 2000). Cocoa butter and alternative fats could distinguished with over 80% accuracy and the type of alternative (cocoa butter versus CBR or CBS) distinguished easily.

Nuclear magnetic resonance (NMR) can be used as a rapid alternative to differential scanning calorimetry in the determination of the solid fat content and studies on the melting behaviour. The determination is based on detection of the different populations of protons in solid and liquid phases, which indicates the hardness of the fat. Hernandez and Rutledge (1994b) used low resolution pulse NMR to compare melting curves of roasted and non-roasted cocoa butters from Africa, Indonesia and South America. Discriminant analysis techniques showed geographical origin to be secondary to the effects of roasting. ^{1}H-NMR and ^{13}C-NMR have both been used with success in combination with multivariate statistics in determining the botanical origin of oils such as those of olive and hazelnut by measurement of small differences in the fatty acid composition (Fauhl *et al.*, 2000). In future, these techniques are likely to be applied more in the authenticity testing of cocoa butters.

Infrared and Raman spectroscopy techniques have also been shown to be of value in determining oil authenticity, particularly when based on the degree of saturation (Aparicio and Baeten, 1998; Bertran *et al.*, 2000), but these methods have not yet been substantially applied to cocoa butter and confectionery fats.

3.4.12 Statistical methods

The advent of relatively inexpensive computers has enabled the accumulation and rapid analysis of large sets of data. By this means patterns and trends not always apparent from visual inspection of chromatograms or tables of data can be discriminated by being sorted into recognizable patterns. This approach is essential for some techniques such as pyrolysis where the quantity of data produced would otherwise be overwhelming. Several statistical approaches to exploit the information content of fatty acid and triacylglycerol patterns for the detection and quantification of CBEs in cocoa butter have been reported (Lipp *et al.*, 2001; Simoneau *et al.*, 1999).

Principal components analysis (PCA) reduces the volume of large data sets by combining correlated variables and maximizing variances to show patterns in the data. Usually, analysis of the variance (ANOVA) is used to prove that the null hypothesis, that there is no difference between the data sets, is not valid. Test results are compared with table values at a probability (normally 95%) that they will conform to that value. Data are plotted in such ways that different populations are visibly separate and the clustering within each set illustrates the degree of repeatability.

Discriminant analysis has been used in many of the analyses described in this chapter, in particular the classification of cocoa butters by origin and processing from pyrolysis MS data (Radovic *et al.*, 1998), from triacylglycerol profiles obtained by HPLC (Hernandez *et al.*, 1991) and from analysis of volatiles (Pino, 1992). Data from the analysis of mixtures of CBEs with cocoa butter, which model techniques for measuring CBEs in chocolate, have been treated by similar means (Anklam *et al.*, 1996).

Multivariate statistical techniques can be used to combine data from several classical analytical methods. In this way Simoneau *et al.* (2000) tested the quantification of CBEs in mixtures with cocoa butter by analysis of fatty acids by GC and triacygylcerols with both GC and HPLC.

3.5 Future issues

Careful selection and breeding of *Theobroma* strains producing butters of preferred properties has been carried out over a long period of time. Genetic modification (GM) uses modern techniques to prevent the expression of genes or to transfer genes from another organism. These techniques could be used in future to benefit the cocoa butter industry by introducing desirable characteristics such as resistance to pests or the production of fat of a more specific composition, or, more likely, an increased yield (Gotsch, 1997). GM is less likely to be applied to CBE producing trees as their growth cycle is very slow. However, it could be applied to oil producing plants, such as rape, in order to modify the oil towards having desirable CBE characteristics. Such oils would therefore have the potential to be used to adulterate cocoa butter, particularly at the chocolate production stage.

In addition many consumers prefer not to consume GM products and expect governments to enforce compositional and labelling standards. Detection of GM of oils presents an analytical problem as GM tests are carried out on proteinaceous material which is not present at high levels in oils and fats, even prior to refining.

The EC Directive of 2000 specifically excludes modifications to the naturally-occurring triacylglycerols of the fat used in CBEs. The widespread use of trans-esterification in the production of CBRs and CBSs means that these materials might become available to those intending to adulterate cocoa butter or ignore the directive. Thus techniques for determining the structure of triacylglycerols are likely to grow in importance.

Future authenticity testing of cocoa butter and chocolate products is likely to follow defined paths. To assure the authenticity of cocoa butters databases of the composition of the relevant fats will be required, constructed from a statistically significantly number of samples of proven provenance and processing history. Analyses must be carried out using internationally validated methods and characterized reference materials by a number of respected laboratories. Scientific co-operation within the European Union is beginning to develop such approaches in ensuring the authenticity of other foods such as olive oil.

Of great importance has been the rapid increase in the use of statistical interpretations of multivariate data made possible by the ready availability of computers. These techniques allow the results from several techniques to be combined, often revealing differences in samples that were not apparent when using the single techniques. The development of statistical methods has been such that the value of classical techniques, such as the comparison of fatty acid or sterol profiles, which can be carried out rapidly with familiar laboratory equipment, has been revived.

Work on cocoa butter and its associated fats, currently prompted by changes in EC legislation, provides a good illustration of concern for authenticity and the skills and inventiveness of analytical chemists. The combined application of established and newer techniques, assisted by new mathematical approaches and the analysis of well-defined samples, provide a paradigm for authenticity testing in general.

References

Ackman, R.G., Hooper, S.N. and Hooper, D.L. (1974) Linolenic artifacts from the deodorization of oils. *J. Am. Oil Chem. Soc.*, **51**, 42–49.

Ando, Y. and Takagi, T. (1993) Micromethod for stereospecific analysis of triacyl-*sn*-glycerols by chiral-phase high performance liquid chromatography. *J. Am. Oil Chem. Soc.*, **70**(10), 1047–1049.

Anklam, E., Lipp, M. and Wagner, B. (1996) HPLC with light scatter detector and chemometric data evaluation for the analysis of cocoa butter and vegetable fats. *Fett Lipid*, **98**(2), 55–59.

Anklam, E., Bassani, E.R., Eiberger, T., Kriebel, S., Lipp, M. and Matissek, R. (1997) Characterisation of cocoa butters and other vegetable fats by pyrolysis mass spectrometry. *Fresenius J. Anal. Chem.*, **357**(7), 981–984.

Aparicio, R. and Baeten, V. (1998) Fats and oils authentication by FT-Raman. *Oléagineux Corps Gras Lipides*, **5**(4), 293–295.

Artaud, J., Dencausse, L., Ntsourankoua, H. and Clamou, J.L. (1995) Comparison of the lipids content of pentadesma and shea butters. *Oléagineux Corps Gras Lipides*, **2**(2), 143–147.

Baliga, B.P. and Shitole, A.D. (1981) Cocoa butter substitutes from mango fat. *J. Am. Oil Chem. Soc.*, **58**(2), 110–114.

Baxter, M., Crews, H. and Anklam, E. (2001) Unpublished data.

Bertran, E., Blanco, M., Coello, J., Iturriaga, H., Maspoch, S. and Montoliu, I. (2000) Near infrared spectrometry and pattern recognition as screening methods for the authentication of virgin olive oils of very close geographical origins. *J. Near Infrared Spectrosc.*, **8**(1), 45–52.

Biino, L. and Carlesi, E. (1971) A research on illipe butter and cocoa butter. *Riv. Ital. Sost. Grasse*, **48**(4), 170–176.

Blicher-Mathiesen, U. (1994) Borneo Illipe, a fat product from different *Shoreas* spp. (Dipterocarpaceae). *Econ. Bot.*, **48**(3), 231–242.

Bracco, U., Rostagno, W. and Egli, R.H. (1967) A new application of the differential thermal analysis for the examination of cocoa butter-1. *Int. Chocolate Rev.*, **22**, 392–396.

Bracco, U., Rostagno, W. and Egli, R.H. (1970) A study of cocoa butter-illipe butter mixtures. *Int. Chocolate Rev.*, **25**, 41–48.

Bruschweiler, H. and Dieffenbacher, A. (1991) Determination of monoglycerides and diglycerides by capillary gas-chromatography—results of a collaborative study and the standardized method. *Pure Appl. Chem.*, **63**(8), 1156–1162.

Buchgraber, M., Ulberth, F. and Anklam, E. (2000) Comparison of HPLC and GLC techniques for the determination of the triglyceride profile of cocoa butter. *J. Agric. Food Chem.*, **48**, 3359–3363.

Byrdwell, W.C. and Emken, E.A. (1995) Analysis of triglycerides using atmospheric-pressure chemical ionization mass-spectrometry. *Lipids*, **30**(2), 173–175.

Carlin, J.T., Lee, K., Hseih, O., Hwang, L., Ho, C.T. and Chang, S. (1986) Comparison of acidic and basic volatile compounds of cocoa butters from roasted and unroasted cocoa beans. *J. Am. Oil Chem. Soc.*, **63**(8), 1031–1036.

Chalseri, S. and Dimick, P.S. (1987) Cocoa butter—its composition and properties. *Manuf. Confect.*, **67**, 115–122.

Chaudhuri, P.G., Chakrabarty, M.M. and Bhattacharyya, D.K. (1983) Modification of some tree borne seed fats for the preparation of high priced confectionery fats. *Fette Seif. Anstrichm.*, **85**(6), 224–227.

Chevalley, J., Rostagno, W. and Egli, R.H. (1970) A study of the physical properties of chocolate. *Int. Chocolate Rev.*, **25**, 3–6.

Christie, W.W. (1995) Silver ion high-performance liquid chromatography, in *New Trends in Lipid and Lipoprotein Analysis* (eds J.-L. Sebedio and E.G. Perkins), AOAC Press, Champaign, Illinois, USA, pp. 59–74.

Codex Alimentarius Commission (1981) Codex standard for cocoa butters. Codex Stan 86-1981.

Codex Alimentarius Commission (2001) Report of the nineteenth session of the Codex Committee on Cocoa Products and Chocolate. Alinorm 03/14.

Coe, S.D. and Coe, M.D. (1996) *The True History of Chocolate*, Thames and Hudson, London.

Coors, V.U. (1991) Utilisation of the tocopherol pattern for recognition of fat and oil adulterations. *Fat Sci. Technol.*, **93**, 519–526.

Crews, C., Calvet-Sarret, R. and Brereton, P. (1997) The analysis of sterol degradation products to detect vegetable fats in chocolate. *J. Am. Oil Chem. Soc.*, **74**(10), 1273–1280.

Crews, C., Calvet-Sarret, R. and Brereton, P. (1999) Steroidal hydrocarbons in refined confectionery fats. *Journal of Chromatogr. A*, **847**(1–2), 179–185.

Dand, R. (1993) *The International Cocoa Trade*. Woodhead Publishing Ltd., Cambridge, UK.

Dencausse, L., Ntsourankoua, H. and Artaud, J. (1995) Comparison of the lipid composition of pentadesma and shea butters. *Oléagineux Corps Gras Lipides*, **2**(2), 143–147.

Deotale, M.Y., Patil, M.N. and Adinarayaniah, C.L. (1990) Thin layer chromatographic detection of kokum butter in cocoa butter. *J. Food Sci. Technol. (Mysore)*, **27**, 230.

Derbesy, M. and Richert, M.T. (1979) Detection of shea butter cocoa butter. *Oléagineux*, **34**, 405–409.

Duchateau, G.S.M.J.E., van Oosten, H.J. and Vasconcellos, M.A. (1996) Analysis of *cis*- and *trans*-fatty acid isomers in hydrogenated and refined vegetable oils by capillary gas-liquid chromatography. *J. Am. Oil Chem. Soc.*, **73**(3), 275–282.

Eder, K. (1995) Gas chromatography of fatty acid methyl esters. *J. Chromatogr. B*, **671**, 113–131.

Erickson, J.A., Weissberger, W. and Keeney, P.G. (1973) Tocopherols in the unsaponifiable fraction of cocoa lipids. *J. Food Sci.*, **38**, 1158–1161.

European Community (2000) EC Directive 2000/36/EC of the European Parliament and of the Council of 23 June 2000 relating to cocoa and chocolate products intended for human consumption.

European Fair Trade Association (1997) 'A clone is being proposed'. It would be false chocolate. Revision of Directive 73/241/EEC Relating to Cocoa and Chocolate Products. EFTA Campaigns Office, Brussels, Belgium (http://www.eftafairtrade.org/pdf/STUDY-EN.DOC).

Fauhl, C., Reniero, F. and Guillou, C. (2000) ^{1}H NMR as a tool for the analysis of mixtures of virgin olive oil with oils of different botanical origin. *Magn. Reson. Chem.*, **38**, 436–443.

Fincke, A. (1975) Detection of shea fat in cocoa butter and cocoa butter substitutes. 1. Detection by Fitelson's reaction. *Deut. Lebensm.-Rundsch.*, **71**(8), 284–286.

Fincke, A. (1976) Cocoa butter and substitute fats—Chemistry and analysis (Kakaobutter und Ersatzfette—chemie und analytik). *Deut. Lebensm.-Rundsch.*, **72**(1), 6–12.

Fincke, A. (1980) Possibilities and limits of gas chromatographic triglyceride analyses for detection of extraneous fats in cocoa butter and chocolate fats. II. Distribution of triglycerides classified by C number in cocoa butter substitutes and other fats. *Deut. Lebensm.-Rundsch.*, **76**(6), 187–192.

Fincke, A. (1982) Possibilities and limits of gas chromatographic triglyceride analyses for detection of extraneous fats in cocoa butter and chocolate fats. IV. Evaluation of gas chromatographic triglyceride analyses of milk chocolate fats. *Deut. Lebensm.-Rundsch.*, **78**(11), 389–396.

Firestone, D. (1994) Direct graphite furnace–atomic absorption method for determination of lead in edible oils and fats: Summary of collaborative study. *J. Assoc. Off. Anal. Chem.*, **77**(4), 951–954.

FSA (Food Standards Agency) (2001) The estimation of cocoa butter equivalents in cocoa butter and chocolate fats by triglyceride analysis: collaborative trial. *Information Bulletin*, No. 13, October 2001.

Frega, N., Bocci, F., Giovannoni, G. and Lercker, G. (1993) High resolution GC of unsaponifiable matter and sterol fraction in vegetable oils. *Chromatographia*, **36**, 215–217.

Garcia Olmedo, R. and Diaz Marquina, A. (1974) Sterols in the unsaponifiable fraction of cocoa butter and chocolate. *An. Bromat.*, **26**(3), 211–239.

Gaydou, E.M. and Bouchet, P. (1984) Sterols, methyl sterols, triterpene alcohols and fatty acids of the kernel fat of different Malagasy mango (*Mangifers indica*) varieties. *J. Am. Oil Chem. Soc.*, **61**(10), 1589–1593.

Geeraert, E. and de Schepper, D. (1983) Structure elucidation of triglycerides by chromatographic techniques. Part 2: RP HPLC of triglycerides and brominated triglycerides. *J. High Res. Chromatogr.*, **6**, 123–132.

Gegiou, D. and Staphylakis, K. (1985) Detection of cocoa butter equivalents in chocolate. *J. Am. Oil Chem. Soc.*, **62**(6), 1047–1051.

Gordon, M.H. and Griffith, R.E. (1992) A comparison of the steryl esters of coconut and palm kernel oils. *Fat Sci. Technol.*, **94**(6), 218–221.

Gotsch, N. (1997) Cocoa biotechnology: Status, constraints and future prospects. *Biotechnol. Adv.*, **15**(2), 333–352.

Grob, K., Giuffré, A.M., Biedermann, M. and Bronz, M. (1994) The detection of adulteration with desterolized oils. *Fat Sci. Technol.*, **96**(9), 341–345.

Gunstone, F. (1999) *Fatty Acid and Lipid Chemistry*, Aspen Publishers Inc., Maryland, USA.

Gunstone, F.D., Harwood, J.L. and Padley, F.B. (1994) *The Lipid Handbook*, 2nd edn. Chapman & Hall, London.

Harwood, J.L. (1991) Cocoa butter, food of the gods? *Chemistry and Industry*, 21 October, pp. 753–756.

Hernandez, C.V. and Rutledge, D.N. (1994a) Multivariate statistical analysis of gas chromatograms to differentiate cocoa masses by geographical origin and roasting conditions. *Analyst*, **119**, 1171–1176.

Hernandez, C.V. and Rutledge, D.N. (1994b) Characterisation of cocoa masses: low resolution pulse NMR study of the effects of geographical origin and roasting on fluidification. *Food Chem.*, **49**, 83–93.

Hernandez, B., Castelloete, A.I. and Permanyer, J.J. (1991) Triglyceride analysis of cocoa beans from different geographical origins. *Food Chem.*, **41**, 269–276.

Hernandez, C., Bermond, A. and Ducauze, C.J. (1994) Using chemometric data to discriminate cocoa masses: analysis of metal contents applied to the determination of their geographical origin and process effect. *Analusis*, **22**, 15–22.

Héron, S. and Tchapla, A. (1994) Choice of stationary phases for separation of mixed triglycerides by liquid phase chromatography. *Analusis*, **22**, 114–126.

Homberg, E. and Bielefeld, B. (1982) Sterols and methylsterols in cocoa butter and cocoa butter substitutes. *Deut. Lebensm.-Rundsch.*, **78**(3), 73–77.

Homberg, V.E. and Bielefeld, B. (1989) Composition and content of sterols in 41 different vegetable and animal fats. *Fat Sci. Technol.*, **91**(1), 23–27.

Itabashi, Y. and Takagi, T. (1986) High performance liquid chromatographic separation of mono-acylglycerol enantiomers on a chiral stationary phase. *Lipids*, **21**, 413–416.

Itabashi, Y., Marai, L. and Kuksis, A. (1991) Identification of natural diacylglycerols as the 3,5-dinitrophenylurethanes by chiral phase liquid-chromatography with mass-spectrometry. *Lipids*, **26**(11), 951–956.

Itoh, T., Tamura, T. and Matsumoto, T. (1974) Sterols and methylsterols in some tropical and subtropical vegetable oils. *Oléagineux*, **29**(5), 253–258.

Itoh, T., Uetsuki, T., Tamura, T. and Matsumoto, T. (1979) Characterization of triterpene alcohols of seed oils from some species of Theaceae, Phytolaccaceaea and Sapotaceae. *Lipids*, **15**(6), 407–411.

Jee, M.H. and Ritchie, A.S. (1984) Two-dimensional thin-layer chromatography technique for use in lipid analysis. *Journal of Chromatogr.*, **299**, 460–464.

Jeffrey, B.S.J. (1991) Silver complexation liquid chromatography for fast high-resolution separation of triacyglycerols. *J. Am. Oil Chem. Soc.*, **68**(5), 289–293.

Kamm, W., Dionisi, F., Fay, L.B., Hischenhuber, C., Schmarr, H.G. and Engel, K.H. (2001) Analysis of steryl esters in cocoa butter by on-line liquid chromatography–gas chromatography. *J. Am. Oil Chem. Soc.*, **918**(2), 341–349.

Kershaw, S.J. (1987) Heterogeneity in commercial contractural samples of illipe nut (*Shorea* spp.) *Int. J. Food Sci. Technol.*, **22**, 67–72.

Kershaw, S.J. and Hardwick J.F. (1981) Heterogeneity in commercial contract analysis samples of shea-nut kernels. *J. Am. Oil Chem. Soc.*, **58**, 706–710.

Kerti, K. (2000) Investigating isothermal DSC method to distinguish between cocoa butter and cocoa butter alternatives. *J. Therm. Anal. Calorim.*, **63**, 205–219.

Kleinert, J. (1964) Pressed and extracted cocoa butter. *Int. Chocolate Rev.*, **19**, 142–153.

Kohiyama, M., Kanematsu, H. and Niiya, I. (1992) Heavy-metals, particularly nickel, contained in cacao beans and chocolate. *J. Jap. Soc. Food Sci.*, **39**(7), 596–600.

Laird, S.A., Obialor, C. and Skinner, E.A. (1996) *An introductory handbook to cocoa certification. A feasibility study and regional profile of West Africa*, Rainforest Alliance, New York.

Lipp, M. and Anklam, E. (1998a) Review of cocoa butter and alternative fats for use in chocolate Part A. Compositional data. *Food Chem.*, **62**, 73–97.

Lipp, M. and Anklam, E. (1998b) Review of cocoa butter and alternative fats for use in chocolate Part B. Analytical approaches for identification and determination. *Food Chem.*, **62**(1), 99–108.

Lipp, M., Simoneau, C. and Anklam, E. (1996) Analysis of triglycerides from cocoa butter, vegetable fats, and their mixtures. Comparison of methods including high pressure chromatography, in *18th International Symposium on Capillary Chromatography, Volume 1* (eds P. Sandra and G. Devos), Huthig (John Wiley), New York.

Lipp, M., Simoneau, C., Ulberth, F., Anklam, E., Crews, C., Brereton, P., de Greyt, W., Schwack, W. and Wiedmaier, C. (2001) Composition of genuine cocoa butter and cocoa butter equivalents. *J. Food Compos. Anal.*, **14**, 399–408.

Liu, J., Lee, T., Bobik, E., Guzmanharty, M. and Hastilow, C. (1993) Quantitative determination of monoglycerides and diglycerides by high-performance liquid chromatography and evaporative light-scattering detection. *J. Am. Oil Chem. Soc.*, **70**(4), 343–347.

Macarthur, R., Crews, C. and Brereton, P. (2000) An improved method for the measurement of added vegetable fats in chocolate. *Food Addit. Contam.*, **17**(8), 653–664.

Macrae, A.R. (1983) Lipid catalysed transesterification of oils and fats. *J. Am. Oil Chem. Soc.*, **60**(2), 291–294.

Md Ali, A.R. and Dimick, P.S. (1994) Thermal analysis of palm mid fraction, cocoa butter and milk fraction blends by differential scanning calorimetry. *J. Am. Oil Chem. Soc.*, **71**(3), 299–302.

Minim, V.P.R. and Cecchi, H.M. (1999) Methodology for the determination of cocoa butter replacements in chocolate by triacylglycerol GC analysis. *HRC-J. High Res. Chromatogr.*, **22**(5), 305–307.

Morales, M.T. and Aparicio, R. (1993) Characterising some European olive oil varieties by volatiles using statistical tools. *Grasas Aceites*, **44**(2), 113–115.

Morales, M.T., Aparicio, R. and Rios, J.J. (1994) Dynamic headspace gas chromatographic method for determining volatiles in virgin olive oil. *J. Chromatogr. A*, **668**, 455–462.

Neri, A., Simonetti, M.S., Cossignani, L. and Damiani, P. (1998) Identification of cocoa butter equivalents added to cocoa butter-I. An approach by fatty acid composition of the triacylglycerol sub-fractions separated by Ag+-HPLC. *Z. Lebensm. Unters. Forsch. A.*, **206**(6), 387–392.

Neri, A., Simonetti, M.S., Cossignani, L. and Damiani, P. (1999) Identification of cocoa butter equivalents added to cocoa butter by fatty acid composition of the triacylglycerol sub-fractions separated by Ag+-HPLC-II. *Z. Lebensm. Unters. Forsch. A.*, **208**(3), 198–202.

Nesareretnam, K. and Razak bin Mohd Ali, A. (1992) Engkabang (illipe)—an excellent component for cocoa butter equivalent fat. *J. Sci. Food Agri.*, **60**, 15–20.

Padley, F.B. and Timms, R.E., (1980) The determination of cocoa butter equivalents in chocolate. *J. Am. Oil Chem. Soc.*, **57**, 286–293.

Peers, K.E. (1977) The non-glyceride saponifiables of Shea butter. *J. Sci. Food Agric.*, **28**, 1000–1009.

Pino, J. (1992) Headspace methods for volatile components of cocoa butter. *Nahrung*, **36**(2), 175–180.

Pino, J. and Roncal, E. (1992) Linalool content in roasted cocoa butter as a characteristic of several flavour grade cocoas. *Nahrung*, **36**(2), 299–304.

Pino, J., Nuñez de Villavicencio, M. and Roncal, E. (1993) Pattern recognition of GC profiles for classification of cocoa butter of Ghanaian and Cuban varieties. *J. Food Quality*, **16**, 125–132.

Podlaha, O., Toeregard, B. and Pueschl, B. (1984) TG-type composition of 28 cocoa butters and correlation between some of the TG-type components. *Lebensm. Wiss. Technol.*, **17**(2), 77–81.

Podlaha, O., Petersson, B. and Toeregard, B. (1985) TG-type composition and some physical characteristic of two kokum butters. *Rev. Fr. Corps Gras*, **32**(5), 201–204.

Radovic, B.S., Lipp, M. and Anklam, E. (1998) Classification of cocoa butters using pyrolysis mass spectrometry. *Rapid Commun. Mass Spectrom.*, **12**(12), 783–789.

Řezanka, T. and Mareš, P. (1991) Determination of plant triacylglycerols using capillary gas chromatography, high performance liquid chromatography and mass spectrometry. *J. Chromatogr.*, **542**, 145–159.

Robinson, J.L. and Macrae, R. (1984) Comparison of detection systems for the high-performance liquid chromatographic analysis of complex triglyceride mixtures. *J. Chromatogr.*, **303**, 386–390.

Robinson, J.L., Tsimidou, M. and Macrae, R. (1985) Evaluation of the mass detector for quantitative detection of triglycerides and fatty acid methyl esters. *J. Chromatogr.*, **324**, 35–51.

Ruiz Méndez, M.V. and Huesa Lope, J. (1991) La manteca de karité. *Grasas Aceites*, **42**(2), 151–154.

Shantha, N.C. and Napolitano, G.E. (1992) Gas chromatography of fatty acids, *J. Chromatogr.*, **624**, 37–51.

Shukla, V.K.S. (1995) Cocoa butter properties and quality. *Lipid Technol.*, May, 54–57.

Shukla, V.K.S. (1997) Chocolate—the chemistry of pleasure. *Inform*, **8**(2), 152–162.

Shukla, V.K.S., Schiotz-Nielsen, W.S. and Batsberg, W. (1983) A simple and direct procedure for the evaluation of triglyceride composition of cocoa butters by high performance liquid chromatography—a comparison with the existing TLC–GLC method. *Fette Seif. Anstrichm.*, **85**(7), 274–278.

Siew, W.L. and Ng, W.L. (1995) Diglyceride content and composition as indicators of palm oil quality. *J. Sci. Food Agr.*, **69**(1), 73–79.

Simoneau, C., Hannaert, P. and Anklam, A. (1999) Detection and quantification of cocoa butter equivalents in chocolate model systems: analysis of triglyceride profiles by high resolution GC. *Food Chem.*, **65**, 111–116.

Simoneau, C., Lipp, M., Ulberth, U. and Anklam, E. (2000) Quantification of cocoa butter equivalents in mixtures with cocoa butter by chromatographic methods and multivariate data evaluation. *Eur. Food Res. Technol.*, **211**(2), 147–152.

Soulier, P., Farines, M. and Soulier, J. (1990) Triterpene alcohols, 4-methylsterols and 4-desmethylsterols of sal and illipe butters. *J. Am. Oil Chem. Soc.*, **67**(6), 388–393.

Staphylakis, K. and Gegiou, D. (1985a) Sterols in cocoa butter. *Fat Sci. Tech.*, **87**(4), 150–155.

Staphylakis, K. and Gegiou, D. (1985b) Free, esterified and glucosidic sterols in cocoa butter. *Lipids*, **20**(11), 723–728.

Takagi, T. and Ando, Y. (1995) Stereospecific analysis of monounsaturated triacylglycerols in cocoa butter. *J. Am. Oil Chem. Soc.*, **72**(10), 1203–1206.

Takahashi, Y., Itabashi, Y., Suzuki, M. and Kuksis, A. (2001) Determination of stereochemical configuration of the glycerol moieties in glycoglycerolipids by chiral phase high-performance liquid chromatography. *Lipids*, **36**(7), 741–748.

Thippesawamy, H.T. and Raina, P.L. (1989) Lipid composition of Kokum (*Garcina indica*) and Dhupa (*Veteria indica*). *J. Food Sci. Technol.* (Mysore), **26**(6), 322–326.

Timms, R.E. and Stewart, I.M. (1999) Cocoa butter, a unique vegetable fat. *Lipid Technol. News.*, October 1999, pp. 101–107.

Traitler, H. and Dieffenbacher, A. (1985) Palm and palm kernel oil in food products. *J. Am. Oil Chem. Soc.*, **62**(2), 417–421.

Von Klagge, P. and Gupta, S.A.K. (1990) Cocoa butter and the equivalents for chocolate products. *Fat Sci. Technol.*, **92**(12), 485–492.

Wong Soon (1991) *Speciality Fats Versus Cocoa Butter*, Kennedy's Publications, London.

Yella Reddy, S. and Prabhakar, J.V. (1989) Confectionery fats from sal (*Shorea robusta*) fat and Phulwara (*Madhuca butyracea*) butter. *Food Chem.*, **34**, 131–139.

Yella Reddy, S. and Prabhakar, J.V. (1990) Cocoa butter substitutes from sal (*Shorea robusta*) fat. *Int. J. Food Sci. Technol.*, **25**, 711–717.

Yella Reddy, S. and Prabhakar, J.V. (1994) Cocoa butter extenders from Kokum (*Garcinia indica*) and Phulwara (*Madhuca butyracea*) butter. *J. Am. Oil Chem. Soc.*, **71**(2), 217–219.

Young, C.C. (1984) The interpretation of GLC triglyceride data for the determination of cocoa butter equivalents in chocolate: A new approach. *J. Am. Oil Chem. Soc.*, **61**(3), 576–581.

4 Authentication of evening primrose, borage and fish oils

N.A. Michael Eskin

4.1 Introduction

Authenticity of oils is important to producers, consumers and regulators. While adulteration does not necessarily pose a health hazard to consumers, it does violate their fundamental rights by selling them a product that is not what it is claimed to be. Such fraudulent practices result when a manufacture replaces, completely or in part, a genuine or more expensive product with a cheaper product. Such adulteration was common in Roman times where legislation was introduced to prevent the fraudulent co-mingling of olive oil. While blending edible oils is a common practice in the preparation of many products, when it differs from the proportions listed on the label or is sold as the genuine item, a fraudulent act has been committed. One of the most commonly cited cases of fraudulent co-mingling was the addition of Malaysian palm olein to cottonseed oil (Rossell, 1991). The addition of borage oil to primrose oil is another potential fraudulent co-mingling reported (Lawson and Hughes, 1988). To prevent such fraudulent practices requires a complete understanding of the composition of the oils and the natural variation of their major constituents. Applying accurate and reliable analytical methods in combination with good statistical procedures will prevent any fraudulent practices by identifying any significant changes in the variability of key constituents in the oil.

The key constituents in oils will vary depending on the particular oil being examined. For example, fatty acid analysis by gas chromatography (GC) is routinely carried out to check the purity of particular oils. In other cases the intact triacylglycerols can be measured using either high temperature gas–liquid chromatography or reversed-phase high performance liquid chromatography (HPLC). Alternatively the analysis of minor components in the unsaponifiable fraction can be an extremely effective way of detecting adulteration. This chapter will discuss the authenticity of borage, primrose and selected fish oils by providing a detailed examination of their respective compositions.

4.2 Fatty acid composition

The one common characteristic that defines the composition of borage, primrose and fish oils is the high degree of unsaturation of their long chain fatty acids. In

Table 4.1 Typical fatty acid composition of evening primrose and borage oils

	Percent of total fatty acids	
Fatty acid	Evening primrose	Borage
C16:0	5.8	10.2
C18:0	2.1	3.3
C18:1	6.6	14.8
C18:2n–6	71.6	37.9
C18:3n–6	12.6	24.6
C18:3n–3	0.2	0.2
C20:0	0.3	0.2
C20:1	0.2	0.2

From Clough (2001).

the case of borage and evening primrose oils it is the presence of high levels of n–6 polyunsaturated fatty acids including α-linoleic (C18:2n–6) and γ-linolenic (C18:3n–6) acids (Table 4.1). However, it is the high level of γ-linolenic acid that differentiates borage and evening primrose oils from most other plant oil sources. In the case of fish oils, they are very rich sources of n–3 polyunsaturated fatty acids, eicosapentaenoic (EPA, C20:5n–3) and docosahexaenoic (DHA, C22:6n–3) acids. These fatty acids are not only nutritionally important but evidence suggests they provide significant health benefits. It is because of these benefits that their authenticity is important, particularly when consumers buy these oils or products based on these oils for their nutritional and health-related properties.

4.2.1 γ-Linolenic acid

γ-Linolenic acid (GLA, C18:3n–6) can theoretically be produced *de novo* in humans by desaturation of linoleic acid as an intermediate in the production of arachidonic acid (Figure 4.1). However, evidence in humans suggests that the conversion of dietary linoleic acid (LA, C18:2n–6) to GLA may be very limited as the enzyme responsible, Δ6-desaturase, is either blocked or fully saturated. As a consequence, the levels of these fatty acids are unaffected by dietary intake of LA but are elevated following consumption of high GLA containing oils such as evening primrose (Gibson and Rassias, 1990). The impairment of Δ6-desaturase has been associated with the effectiveness of oral GLA in the treatment of a number of diseases including premenstrual syndrome (Brush *et al.*, 1984; Ockerman *et al.*, 1986), rheumatoid arthritis (Belch *et al.*, 1988), breast disorders (Mansel *et al.*, 1990) and atopic eczema (Manku *et al.*, 1984). As a result, a number of countries allow the sale of evening primrose oil (EPO) brands in pharmacies and health stores. It is anticipated that products based on borage oil

Linolenic acid (C18:2n–6)

↓

γ-Linolenic acid (C18:3n–6)

↓

Dihomo-γ-linolenic acid (C20:3n–6)

↓

Arachidonic acid (C20:4n–6)

Figure 4.1 Metabolism of n–6-fatty acids.

will find a similar market. Consequently it is important to ensure such products are genuine.

4.2.2 Eicosapentaenoic and docosahexaenoic acids

The n–3 polyunsaturated fatty acid (PUFA) content of a number of sea and freshwater fish are listed in Table 4.2 in decreasing order of fat content. While these are average values it is important to note that the n–3 fatty acid content varies depending on the seasons and regions where the fish is caught. Gamez-Meza *et al.* (1999) reported much higher levels of eicosapentaenoic acid (EPA, C20:5n–3) and docosahexaenoic acid (DHA, C22:6n–3) in sardine oil caught in the Gulf of California. The level of n–3 fatty acids in fish is related to their

Table 4.2 Typical n–3 fatty acid content of sea and freshwater fish

Fish	Fat content (g/100 g)	EPA (g/100 g)	DHA (g/100 g)
Eel	24.5	0.26	0.57
Herring	17.8	2.04	0.68
Sprat	16.6	1.33	1.90
Tuna	15.5	1.08	2.29
Salmon	13.6	0.71	2.15
Mackerel	11.9	0.63	1.12
Carp	4.8	0.22	0.08
Sardine	4.5	0.58	0.81
Swordfish	4.4	0.13	0.66
Red fish	3.6	0.28	0.13
Trout	2.7	0.15	0.44
Halibut	1.7	0.14	0.37
Sole	1.4	0.03	0.16
Pike-Perch	0.7	0.10	0.05
Cod	0.6	0.06	0.12
Haddock	0.6	0.04	0.12

Adapted from Trautwein (2001). The data presented are average values.
EPA, eicosapentaenoic acid; DHA, docosahexaenoic acid.

α-Linolenic acid (ALA, C18:3n–3)

↓

Eicosapentaenoic acid (EPA, C20:5n–3)

↓

Docosahexaenoic acid (DHA, C22:6n–3)

Figure 4.2 Metabolism of n–3 fatty acids.

ability to adapt to cold water temperatures. For example, fish caught in the cooler waters of the North Atlantic are fatter and higher in n–3 fatty acids compared with fish caught in the Baltic sea or more tropical waters. This chapter will confine itself to those fish oils with high levels of EPA and DHA that are being produced commercially or have considerable potential as encapsulated oils.

EPA and DHA in fish is derived primarily from ingesting marine algae where it is synthesized by desaturation and elongation of α-linolenic acid (ALA, C18:3n–3) (Figure 4.2).

In humans, the conversion of ALA to EPA and DHA is extremely slow, with only about 15% and 5% of ALA converted to EPA and DHA, respectively (Cunnane, 1995). This conversion appears to be affected by a number of dietary factors. For example, a diet rich in linoleic acid has been found to reduce this conversion by as much as 40% (Emken, 1995). In addition, saturated and *trans* fatty acids also interfere with ALA desaturation and elongation steps (Ackman and Cunnane, 1992; Houwelingen and Hornstra, 1994). DHA can be reconverted back to EPA, although in humans it appears to be a very minor pathway (Brossard *et al.*, 1996). DHA appears to play an important in the brain and retina and was found to be incorporated during the last trimester of pregnancy and the first year of life. Visual acuity was shown to develop much faster in preterm infants fed formulas rich in DHA compared with standard infant formulas low in long chain n–3 fatty acids (Jorgensen *et al.*, 1996).

4.3 High GLA oils

4.3.1 Evening primrose oil

Evening primrose (*Oenothera biennis* L.), a biennial plant belonging to the Onagraceae family, is considered a weed native to North America. It is cultivated in a number of countries on account of the nutritional and pharmaceutical properties of the oil which is in constant demand (Carter, 1988). The oil extracted, referred to as evening primrose oil (EPO), is a good source of GLA. Unlike commercial oilseeds, evening primrose does not produce a high yield of seeds, but is a preferred source of GLA as it does not contain any ALA.

Table 4.3 Reproducibility data for replicate analyses of evening primrose oil by gas chromatography

	Sample 1		Sample 2[a]	
Fatty acid	%[b]	CV[c](%)	%	CV(%)
C12:0	0.04	7.1	0.03	7.5
C14:0	0.07	4.8	0.07	12.3
C15:0	0.03	7.1	0.03	7.4
C16:0	6.44	1.4	5.93	0.4
C16:1 (n–9)	0.04	7.2	0.04	8.3
C16:1 (n–7)	0.07	9.1	0.04	17.8
C17:0	0.06	9.1	0.07	8.8
C17:1 (n–8)	0.04	6.2	0.02	3.6
C18:0	1.81	0.6	2.08	1.6
C18:1 (n–9)	9.76	0.8	4.99	1.1
C18:1 (n–7)	0.84	1.5	0.56	2.3
C18:2 (n–6)	71.20	0.1	74.40	0.2
C18:3 (n–6)	8.72	0.3	10.80	0.1
C18:3 (n–3)	0.18	3.9	0.17	7.6
C20:0	0.31	9.5	0.31	4.5
C20:1 (n–9)	0.23	10.0	0.23	5.2
C20:2 (n–6)	0.03	3.9	0.05	5.0
C22:0	0.12	10.0	0.13	11.3
C24:0	0.46	4.6	0.06	7.5

Reprinted from Court *et al.* (1993), © 1993, with permission from Elsevier Science.
[a]Mean value of 7 determinations.
[b]Weight per cent of total fatty acids.
[c]Coefficient of variation.

Differences in the minor fatty acids present in EPO have been reported by a number of researchers which could be due, in part, to genetic variations (Botazzi *et al.*, 1985; Gibson *et al.*, 1992; Lotti *et al.*, 1984; Muderhwa *et al.*, 1987; Singer *et al.*, 1990). Court *et al.* (1993) developed a reliable and reproducible method for analysing the fat content and fatty acids in a large number of small seed samples of evening primrose. Polytron extraction was found to be as reliable as extraction using the Soxhlet and could routinely extract 250 mg of seed. Analysis of the fatty acid methyl esters, involved separation and quantitation by capillary chromatography. Acceptable reproducibility was obtained for the majority of fatty acids with the exception of a few very minor fatty acids (Table 4.3).

To further the identification of the minor fatty acids in evening primrose oil, methyl esters of fatty acids were separated by column chromatography on a silver nitrate–silicic acid absorbent according to their degree of unsaturation (De Vries, 1963). The data obtained are summarized in Table 4.4 based on identification by gas chromatography–mass spectrometry. Of the saturated fatty acids, C13:0, C19:0, C21:0 and C23:0 had not been reported previously. A number of monounsaturated fatty acids have been reported previously in EPO

Table 4.4 Fatty acids in evening primrose (*Oenothera biennis* L.) wild biotypes

		Relative retention time		
Peak number	Fatty acid	CPS-1[a]	SP-10[b]	Identification
Saturated fatty acids				
1	C12:0	0.403	0.419	GC–MS, RT
2	C13:0	0.495	0.495	GC–MS, RT
3	C14:0	0.556	0.550	GC–MS, RT
4	C15:0	0.664	0.639	GC–MS, RT
5	C16:0	0.776	0.751	GC–MS, RT
7	C17:0	0.888	0.864	GC–MS, RT
10	C18:0	1.000	1.000	GC–MS, RT
13	C19:0	1.109	1.142	GC–MS, RT
19	C20:0	1.215	1.313	GC–MS, RT
23	C21:0	1.357	1.515	GC–MS, RT
26	C22:0	1.442	1.776	GC–MS, RT
27	C23:0	1.629	2.076	GC–MS, RT
28	C24:0	1.774	2.468	GC–MS, RT
Monounsaturated fatty acids				
4	C14:1	0.564	0.567	GC–MS
6	C15:1	0.670	0.658	GC–MS
8	C16:1 (n–9)	0.793	0.773	GC–MS, RT
9	C16:1 (n–7)	0.802	0.780	GC–MS, RT
11	C17:1 (n–8)	0.910	0.899	GC–MS, RT
14	C18:1 (n–9)	1.021	1.035	GC–MS, RT
15	C18:1 (n–7)	1.029	1.042	GC–MS, RT
24	C20:1 (n–9)	1.234	1.358	GC–MS, RT
Polyunsaturated fatty acids				
12	C17:2	0.966	0.972	GC–MS
16	C18:2	1.052	1.086	GC–MS
17	C18:2 (n–6)	1.080	1.115	GC–MS, RT
18	C18:3 (n–6)	1.101	1.158	GC–MS, RT
20	C18:3 (n–3)	1.131	1.207	GC–MS, RT
21	C18:4 (n–3)	1.159	1.264	GC–MS, RT
25	C20:2 (n–6)	1.283	1.455	GC–MS, RT

Reprinted from Court *et al.* (1993), © 1993, with permission from Elsevier Science.
[a]Quadrex CPS-1 capillary column (25 m × 0.25 mm i.d.).
[b]Supelco Supelcowax 10 capillary column (60 m × 0.25 mm. i.d.).
GC–MS, Gas chromatography–mass spectrometry; RT, retention time.

(Botazzi *et al.*, 1985; Gibson *et al.*, 1992; Lotti *et al.*, 1980). In this study eight monounsaturated fatty acids were readily detected, with the major one being C18:1 with several minor ones, C14:1 and C15:1, only detected in the enriched column fractions. The major C16:1 fatty acids identified in evening primrose were n–7 and n–9, which contrasted with borage oil that contained C16:1 (n–7) and C16:1 (n–5) (Wretensjo and Svensson, 1990).

The presence of C22:1 and C24:1 in commercial EPO detected by Gibson *et al.* (1992) suggested possible adulteration of the oil as these fatty acids were not found in any of the seeds they analysed. However, other studies did report the presence of C22:1 in evening primrose (Muderhwa *et al.*, 1987; Singer *et al.*, 1990). These fatty acids were not detected by Court *et al.* (1993) in their seed samples, with the exception of one sample which had several peaks that were quite negligible and did not warrant further examination.

4.3.2 *Borage oil*

Borage oil (BO), extracted from the seeds of borage (*Borago officinalis* L.), is receiving increased attention because of its high levels of GLA, reported to be between 20% and 25% (Beaubaire and Simon, 1987; Traitler *et al.*, 1984). The oil itself consisted of 95.7% neutral lipids, 2.3% glycolipids and 2.3% phospholipids (Senanayake and Shahidi, 2000). The neutral fraction is predominantly triacylglycerols (TAGs) (99.1%), together with very small amounts of diacylglycerols (DAGs) (0.06%), monoacylglycerols (MAGs) (0.02%) and sterols (0.02%).

The quantitative analysis of complex fatty acids by gas chromatography of their methyl ester derivatives on fused-silica capillary columns of moderate polarity is the standard method of choice. However, methyl esters of fatty acids proved unsuitable for identification by mass spectrometry because the unsaturated fatty acids underwent isomerization during ionization. To stabilize the double bond during ionization different derivatives were formed including picolinyl esters (Christie *et al.*, 1986). Unacceptable poor resolution and peak overlapping of the picolinyl fatty esters, however, was found when analysed by GC with packed columns coated with non-polar stationary phases. Wrestensjo and Svensson (1990) successfully overcame this problem by analysing picolinyl ester derivatives of borage oil fatty acids by capillary GC on a medium-polarity column, which were then subjected to mass spectrometry. The GC elution patterns for methyl esters and picolinyl derivatives for borage oil, in spite of different retention times, were very similar as seen in Figure 4.3. The mass spectra of the picolinyl ester derivatives for borage oil fatty acids produced good molecular ions and distinct diagnostic ions, which permitted identification of the fatty acids as summarized in Table 4.5. Two monoenoic fatty acids isomers of C16:1 were found, n–7 and n–5, neither of which had been reported previously. Using medium-polarity columns, these researchers obtained excellent separation of positional isomers as well as linoleic and α-linolenic acids. In addition, minor fatty acids, C16:1n–5, C18:3n–3 and C18:4n–3, were also present at 0.21, 0.20 and 0.18%, respectively. The accuracy of this method, together with the identification and quantitation of these minor fatty acids, could provide additional evidence for the authenticity of borage oil.

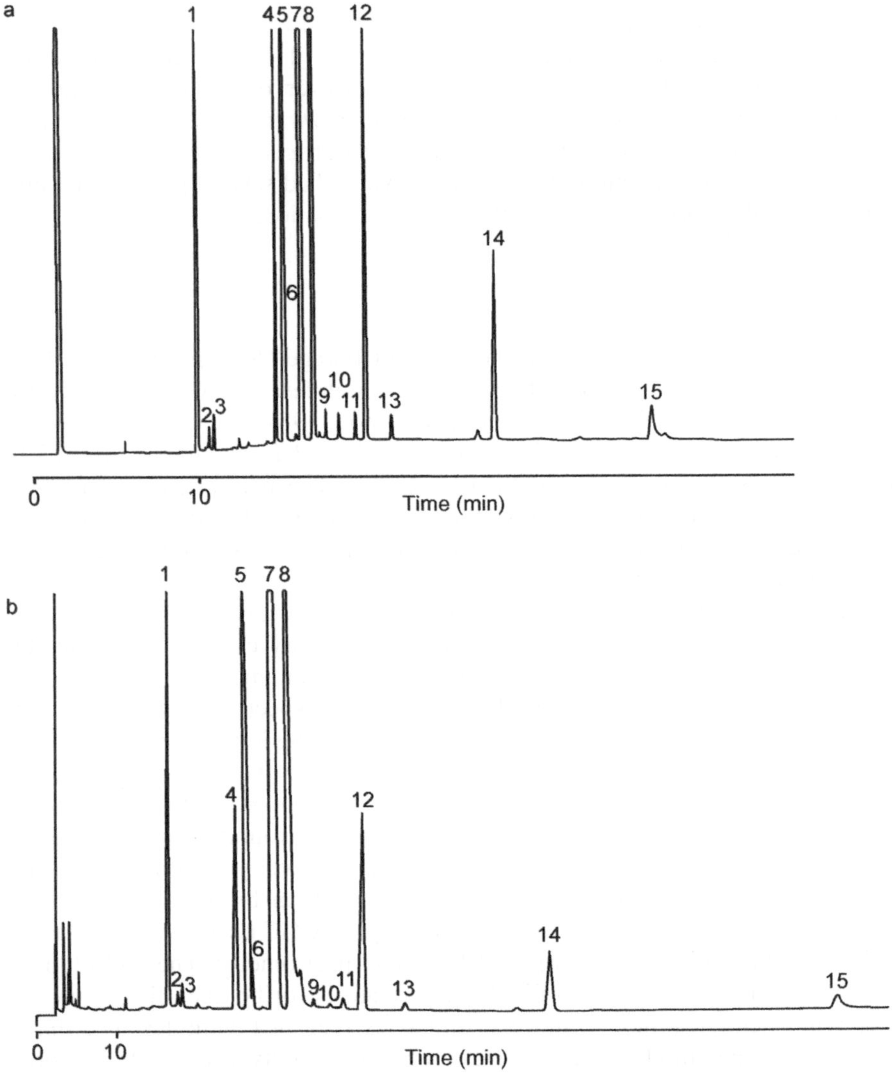

Figure 4.3 Reprinted from Wrestensjo and Svensson (1990), © 1990 with permission from Elsevier Science.

In any accurate analysis, elimination of artefacts is critical as they could interfere with characterizing a particular oil. Geometrical isomers of ALA (*cis*-9, *cis*-12, *cis*-15 18:3) were reported as common constituents of fully refined ALA rich oils as well as in foods containing these oils (Wolff, 1992, 1993a,b; Wolff and Sebedio, 1991). However, a number of researchers suggested these

Table 4.5 GC–MS analysis of fatty acid picolinyl esters of borage oil

Peak number	Component	Retention time (min)	Molecular ion m/z
1	16:0	12.35	347
2	16:1(n–7)	13.35	345
3	16:1(n–5)	13.61	345
4	18:0	19.21	375
5	18:1(n–9)	20.28	373
6	18:1(n–7)	21.63	373
7	18:2(n–6)	22.87	371
8	18:3(n–6)	24.18	369
9	18:3(n–3)	27.43	369
10	18:4(n–3)	28.49	367
11	20:0	30.63	403
12	20:1(n–9)	32.12	401
13	20:2(n–6)	35.67	399
14	22:1(n–9)	45.98	429
15	24:1(n–9)	64.36	457

Adapted from Wretensjo and Svensson (1990).

geometrical isomers formed from the heat treatment (190–200°C or higher) to which the oil is subjected during deodorization (Ackman *et al.*, 1974: Devinat *et al.*, 1980). The possible formation of GLA geometrical isomers in borage oil when heated either under vacuum or during steam-vacuum deodorization was investigated by Wolff and Sebedio (1994). These researchers found the DB-Wax capillary column best suited for studying the behaviour of either methyl or isopropyl esters of this fatty acid.

The two main geometrical isomers of GLA formed during heating were *trans*-6, *cis*-9, *cis*-12 and *cis*-6, *cis*-9, *trans*-12 18:3 acids with minor amounts of *cis*-6, *trans*-9, *cis*-12. The effect of different deodorization temperatures on the fatty acids of borage oil is shown in Table 4.6. Under these conditions, total *trans*-18:3n–6 isomers, completely absent in the native oil, increased to 0.2, 1.2 and 3.9% of the total fatty acids when steam deodorized at 200°C, 220°C and 240°C, respectively. These changes corresponded to a relative increase in the degree of isomerization (DI) of 0.9, 4.6 and 15.6%, respectively. It was apparent from these results that deodorizing borage oil at temperatures below 200°C could minimize the formation of these isomers. However, the presence of these isomers could be a useful marker to indicate whether borage oil was subjected to temperature abuse or to deodorization temperatures of 200°C or higher.

4.3.3 Triacylglycerol structure of EPO and BO

The position of fatty acids in the triacylglycerol (TAG) structure not only significantly influences fatty acid absorption, but could provide important markers for establishing the authenticity of different oils. Lawson and Hughes (1988) determined the TAG stereospecific structure of a number of GLA rich oils,

Table 4.6 Formation of geometrical isomers of γ-linolenic acid in borage oil deodorized at different temperatures as determined by gas liquid chromatography of fatty acid isopropyl esters on a 30 m long DB-wax capillary column

Fatty acid	Native oil[a] (n = 2)[b]	200°C 2 h (n = 2)	220°C 2 h (n = 2)	240°C 2 h (n = 2)
Sum 18:3n–6	25.40	25.27	25.21	24.89
cis, cis, cis-18:3n–6	25.40	25.04	24.05	21.00
trans, cis, cis-18:3n–6	—	0.09	0.53	1.72
cis, cis, trans-18:3n–6	—	0.13	0.63	1.97
cis, trans, cis-18:3n–6[c]	—	trace	trace	0.20

Adapted from Wolff and Sebedio (1994), data given as peak area percentages. [a]refined borage oil not deodorized and used for deodorization experiments; [b]number of analyses; [c]probably includes some *trans*-6, *cis*-9, *trans*-12 18:3 acids. Generally too low to be taken into account by the integrator.

Table 4.7 Triacylglycerol stereospecific analysis of evening primrose and borage oils

		Fatty acid (mol %)				
Oil	Position[a]	16:0	18:0	18:1	18:2n–6	18:3n–6
Evening primrose	All	5.9	1.8	7.5	74.8	9.3
	1	9.6 ± 0.3	4.4 ± 0.1	8.0 ± 0.3	73.3 ± 0.6	3.6 ± 0.2
	2	1.8 ± 0.1	1.5 ± 1.4	7.6 ± 0.6	78.9 ± 0.4	10.7 ± 0.3
	3	6.8 ± 0.6	0.3 ± 0.4	7.1 ± 1.1	74.2 ± 4.8	13.5 ± 0.5
Borage oil	All	10.7	3.0	15.4	38.1	24.8
	1	16.9 ± 1.4	5.3 ± 0.2	13.8 ± 0.4	54.3 ± 0.9	4.0 ± 0.3
	2	2.3 ± 1.2	1.3 ± 1.2	13.7 ± 1.2	42.5 ± 1.3	40.4 ± 2.2
	3	13.1 ± 1.4	3.1 ± 0.4	18.7 ± 1.2	17.6 ± 1.6	30.1 ± 1.9

Adapted from Lawson and Hughes (1988).
[a]Position 1, lysophosphatidylcholine; position 2, 1,2-diacylglycerol × 2 minus lysophosphatidylcholine; position 3, triacylglycerol × 3 minus 1,2-diacylglycerol × 2.

including borage and evening primrose, using different lipases. The fatty acids of most plant oils are symmetrically distributed between the *sn*-1 and *sn*-3 positions with saturated fatty acids preferring the *sn*-1 and *sn*-3 positions and LA the *sn*-2 position. Using a single method for determining the nature of each acyl position, they obtained the results for EPO and BO TAGs shown in Table 4.7.

Palmitic acid was highest in the 1-position for both oils but was still present in significant amounts in the 3-position. Stearic acid was also predominant in the 1-position for both oils but was present in high amounts in the 3-position in borage oil only. Very little GLA was detected in position 1 in either of the oils, while the highest levels were in the 3-position and the 2-position of evening primrose and borage oils, respectively. Linoleic acid showed a small preference at the 2-position in evening primrose oil compared with a preference at the 1-position in borage oil. Oleic acid was evenly distributed among the three positions in evening primrose oil but was higher in the 3-position in borage oil. These subtle differences could be helpful in differentiating between these two plant oils.

Laakso (1997) analysed the TAGs of oils rich in α- and γ-linolenic acid using silver ion HPLC connected to an online atmospheric pressure chemical ionization–mass spectrometric (APCI–MS) detection system. The technique provided important information about the structure of seed oils. The most abundant TAG fractions of borage oil separated were C18:1/C18.2/C18:2 + C18:0/C18:1/C18:3n–6, C18:2/C18:2/C18:2 + C18:0/C18:2/C18:3n–6, C18:1/C18:2/C18:3n–6 and C18:2/C18:2/C18:3n–6. Altogether 79 molecular species of TAGs were identified, reflecting the great complexity of BO. In EPO, the most abundant TAG fractions were C18:0/C18:2/C18:2, C18:1/C18:2/C18:2, C18:2/C18:2/C18:2 + C18:0/C18:2/C18:2n–6 and C18:2/C18:2/C18:3n–6 with 39 TAG molecular species identified.

Bergana and Lee (1996) used ^{13}C nuclear magnetic resonance (NMR) to examine the positional distribution of fatty acids in the glycerol backbones of BO and EPO. Using this technique they were able to resolve the resonance from carbonyl compounds at the 1,3-positions from that at the 2-position. This technique could lead to the development of databases for establishing the positional distribution of TAGs containing medium-chain saturated to long-chain polyunsaturated fatty acids and could be extended to more complex systems such as fish oils. This information would further the identification and authenticity of the particular oil of interest to regulators.

4.3.4 *Unsaponifiable fraction of EPO and BO*

The unsaponifiable fraction of most plant oils is relatively small and includes both sterols and tocopherols. Nevertheless the presence of distinct components in this fraction could provide a useful fingerprint for determining the authenticity or purity of these oils.

4.3.4.1 *Tocopherols*

Uzzan *et al.* (1992) suggested tocopherol content and composition may be good indicators of vegetable oil quality and purity. They analysed the tocopherols in four BO samples (two virgin, two refined), three EPO samples (two virgin, one refined) and one blackcurrant seed oil. Their results (in Table 4.8) showed BO samples were much higher in total tocopherols compared with EPO, as two samples of EPO had added α-tocopherol acetate (samples 6 and 7). A close examination of tocopherol isomers showed the primary difference between borage and evening primrose oils was the predominance of the δ-isomer in the former. In contrast, the major tocopherol isomer in evening primrose was the γ-isomer followed by the α-isomer. α-Tocopherol were present at low levels in BO compared with trace amounts of δ-tocopherol in EPO. Unfortunately, these researchers did not indicate the particular varieties examined or which samples

Table 4.8 Tocopherol content and composition of borage and evening primrose and blackcurrant oils

	Borage oil				Evening primrose oil			Blackcurrant oil
Sample	1	2	3	4	5	6	7	8
Total tocopherols	1104	1111	1045	732	454	12,786	12,502	1043
α-Tocopherol	46	trace	6	9	99	194	356	20
β-Tocopherol	5	—	—	—	—	—	—	8
γ-Tocopherol	272	98	83	33	340	358	298	647
δ-Tocopherol	781	1103	956	690	15	19	10	68
Acetate α-tocopherol	—	—	—	—	—	12,215	11,841	—

Adapted from Uzzan *et al.* (1992); tocopherols determined as mg/kg.

were refined. Later research by Sensidoni *et al.* (1996) confirmed the absence of α-tocopherol in BO. These differences in tocopherol isomers could be useful markers in establishing the purity and authenticity of evening primrose and borage oils.

Blackcurrant seed oil, another high GLA oil (Traitler *et al.*, 1984), is also a rich source of tocopherols (Table 4.8). Goffman and Galletti (2001) recently reported blackcurrant oil contained up to 15.8% GLA. This oil is a potential adulterant of EPO or BO. Uzzan *et al.* (1992) found blackcurrant oil was high in γ- and α-tocopherols compared with EPO and BO (Table 4.8). These isomers were also shown by Goffman and Galletti (1992) to be the primary ones present in blackcurrant oil. The presence of high levels of γ- and α-tocopherols in blackcurrant oil could be useful for detecting its adulteration in EPO and BO.

4.3.4.2 *Phytosterols*

Phytosterols generally represent the major portion of the unsaponifiable fraction and could also serve as another fingerprinting technique for oil authentication. These sterols are present in the free or esterified forms differing from each other primarily in the side chain and the number and location of double bonds in the ring structure (Mannino and Amelotti, 1975; Kochar, 1983). Very little information is available on BO and EPO sterols with the exception of a paper by Reina *et al.* (1997). These researchers compared the relative and absolute compositions of the sterols and triterpene diols in a number of plant oils including evening primrose. Table 4.9 compares the sterol composition between cottonseed and evening primrose oils. Based on the absolute composition data, EPO had a total sterol content of 1039 mg/100 g of oil, which was almost three times that found in cottonseed oil (380 mg/100 g of oil). β-Sitosterol accounted for over 80% of the total sterols in both oils, while campesterol and Δ^5-avenosterol were present in much higher amounts in EPO. These differences, together with the much higher levels of sterols in EPO, could provide further evidence for its authentication. Similar work is needed to identify the sterols in BO to establish any unique sterol pattern that could distinguish it from EPO.

Table 4.9 Sterol composition of evening primrose and cottonseed oils

	Evening primrose		Cottonseed	
Sterol	Absolute composition	Relative% composition	Absolute composition	Relative% composition
Cholesterol	0.62	0.05	1.19	0.31
Campesterol	91.8	8.84	29.3	7.80
Campestanol	3.99	0.38	0.70	0.17
Sigmasterol	0.97	0.10	3.68	0.98
Clerosterol	7.22	0.70	3.20	0.85
β-Sitosterol	864	83.3	325	86.7
Sitostanol	13.2	1.27	2.15	0.57
Δ^5-Avenosterol	48.7	4.68	7.21	1.91
$\Delta^{5,24}$-Stigmastadienol	3.98	0.38	0.41	0.10
Δ^7-Stigmasterol	2.15	0.21	1.64	0.44
Δ^7-Avenosterol	1.88	0.18	0.97	0.25
Erythrodiol	—	—	4.47	1.18

Adapted from Reina *et al.* (1997); each result represents the average of three oils tested; absolute composition calculated for each identifiable sterol as mg/100 g oil; relative% composition calculated as ratio of each sterol peak/total sterol peak area.

Values for it are listed by the AOCS (1997), which indicates that campesterol is high at 25–30%, stigmasterol is absent, and high levels of Δ^5-avenosterol and 24-methylene cholesterol are present; however, a source is not given for these values and they should be confirmed. This publication also lists blackcurrant seed oil as containing: 7.2–10.4% campesterol; 0.5–1.0% stigmasterol; 70–85% β-sitosterol; 2–5% Δ^5-avenosterol; 0.5–4.5% Δ^7-stigmasterol and 0.4–2% Δ^7-avenasterol.

4.4 Fish oils

4.4.1 Sardine oil

Fish oils, important sources of n–3 polyunsaturated fatty acids, are a major by-product of the fish industry. While a particular fish species can be identified by DNA analysis using polymerase chain reaction–restriction fragment length polymorphism techniques (Rehbein *et al.*, 1997; Carrera *et al.*, 1999), authentication of the corresponding oil is more difficult. The consumption of hydrogenated fish oil in the USA during 1921–1951 stopped with the failure of the sardine fishery in California in the late 1940s (Bimbo, 1989; Marin and Flick, 1990). The resurgence of the sardine fishery in 1967 produced 322,000 tons in 1988–1989 which made sardines (*Sardinops sagax caeruleus*) an important commercial species in northwest Mexico. These fish are used primarily for animal feeds, although their n–3 rich oil is considered a valuable by-product. However, one of the problems associated with fish and fish products is variability

Table 4.10 Polyunsaturated fatty acid composition of oils from whole sardine (*Sardinops sagax caeruleus*) from the Gulf of California, Mexico

	Total fatty acids (wt%)			
Fatty acids	February	April	June	Average
C18:2n–6	2.07 ± 0.38	1.16 ± 0.47	1.60 ± 0.11	1.61
C18:3n–3	0.39 ± 0.10	0.48 ± 0.26	0.47 ± 0.08	0.44
C18:4n–3	2.31 ± 0.10	2.00 ± 0.57	1.79 ± 0.21	2.03
C20:4n–6	0.92 ± 0.39	0.91 ± 0.25	1.11 ± 0.14	0.98
C20:5n–3	23.91 ± 0.98^{a}	19.21 ± 0.95^{b}	18.26 ± 1.30^{b}	20.46
C22:6n–3	9.61 ± 1.69^{c}	13.70 ± 2.28^{d}	13.43 ± 2.06^{c}	12.24
Σ Polyunsaturates	39.21 ± 2.03^{e}	37.46 ± 3.39^{e}	36.66 ± 1.25^{e}	37.76

Adapted from Gamez-Meza *et al.* (1999).
Values in rows with different superscripts (a–e) are significantly different ($P < 005$).

due to season and types of feed (Stansby, 1981; Krzynowek *et al.*, 1992). This variability in fatty acid composition could make it difficult to establish the authenticity of the particular fish oil.

Gamez-Meza *et al.* (1999) examined the effect of seasonal variation on the fatty acid composition of sardine oil. Their results in Table 4.10 summarize the variations in the polyunsaturates in sardine oil from catches obtained in the Gulf of California between February and June of 1994/95. The total polyunsaturates accounted for 37.8% of the total fatty acids, with the most abundant being EPA and DHA, with a mean content of 20.4% and 12.2%, respectively. While total polyunsaturates did not differ significantly there were some differences in EPA and DHA levels. EPA was significantly higher in oil collected in February compared with April and June. The opposite was true for DHA levels which were significantly lower in the February catch compared with April and June. This probably reflected differences in the availability of phytoplankton during the early and late seasons and was attributed to the spawning cycle (Lluch-Belda *et al.*, 1986). Similar variations were reported previously by Bandarra *et al.* (1997) for sardine species caught over a 12-month period off the Portuguese coast. Gamez-Meza *et al.* (1999) reported that the total n–3 content of their sardine oil (34.10–36.26%) was similar to a high n–3 commercial encapsulated fish oil product (38.34%) and significantly higher than another encapsulated fish oil product (28.66%). These researchers suggested that refined sardine oil had considerable potential as an encapsulated product or as an ingredient in foodstuffs.

4.4.2 Menhaden oil

Besides sardines, menhaden (*Brevoortia tyrannus*), a member of the herring family, is the mainstay of the USA Atlantic Coast and Gulf of Mexico fisheries. The oil from menhaden was consumed in Europe as a food oil for well over 60

Table 4.11 Fatty acid composition of menhaden oil

Fatty acid	Weight (%)
C14:0	10.7
C14:1	0.4
C16:0	20.6
C16:1	14.1
C18:0	3.8
C18:1	9.8
C18:2n–6	2.2
C18:3n–3	2.5
C20:4n–6	2.0
C20:5n–3	19.2
C22:5n–3	3.5
C22:6n–3	11.2

Adapted from Lee and Foglia (2001).

years and commercially processed for almost a century and a quarter. The fatty acid composition of menhaden oil is shown in Table 4.11. It is quite high in n–3 fatty acids and received 'generally regarded as safe' (GRAS) approval by the USA Foods and Drugs Administration in 1989 for use in margarines. Today, refined menhaden oil is readily available commercially as an encapsulated oil marketed for its high levels of n–3 fatty acids.

4.4.3 Encapsulated fish oils

The popularity and availability of Brazilian encapsulated fish oils in health food stores, pharmacies and supermarkets is associated with the health benefits derived from n–3 fatty acids. These oils, rich in polyunsaturated fatty acids, are extremely susceptible to oxidation with the formation of undesirable oxidized polymers. Fantoni *et al.* (1996) analysed 16 trademarked encapsulated fish and cod liver oils sold in Brazil and found the quality varied with peroxide values ranging from 2.1 to 20.3 meq/kg of the oil. The cut-off point for oil quality is a peroxide value of 5 meq/kg, established by the *Codex Alimentarius* (Mounts, 1994), indicating 7 out of 16 encapsulated oils were outside these specifications. However, the polar components for these same samples ranged from 0.1 to 8.3%, which were all well below the maximum value of 25% permitted for a frying oil. Fatty acid analyses of the encapsulated oils indicated the majority were marine oils, although the levels of EPA and DHA ranged from 0.5–18.9% and 0.6–13.2%, respectively. Some manufacturers may fraudulently add vegetable oils to reduce the level of polyunsaturated fatty acids, to minimize oxidative changes and reduce costs. This practice also reduces any health benefits associated with these marine oils. However, only one sample had less than 1% EPA and DHA. Further examination of the five encapsulated oil samples with labels listing their EPA + DHA content proved to be authentic as the analytical data closely

Table 4.12 Comparison of EPA + DHA claims and actual measurements for encapsulated fish oils

Sample	Label claims (mg/g EPA + DHA)	Analysed samples (mg/g EPA + DHA)
1	300	279
2	300	321.4
3	300	287.6/305.0
7	300	235.1
8	300	312.4

Adapted from Fantoni *et al.* (1996).
EPA, eicosapentaenoic acid (C20:5n–3); DHA, docosahexaenoic acid (C22:6n–3).

matched the stated claims (Table 4.12). These researchers pointed out that the encapsulated cod liver oil samples tended to be lower in EPA + DHA as they were designed primarily to provide vitamins A and D. An earlier study by Shukla and Perkins (1991) found five out of six encapsulated fish oil products sold in the USA contained oxidized dimeric TAG polymers (1–10%) and one sample contained trimeric (6.3%) and oligomeric (3.1%) TAG polymers. The latter sample also contain 36.3% total polars exceeding the maximum level of 25% total polars permitted for frying oils in Europe.

Because of the toxic nature of these oxidized components, the authors cautioned regular intake of encapsulated fish oils. The sale of oxidized fish oil products can be considered criminal as they pose potential health risks to the consumer. In addition to ensuring that the consumer is buying the authentic product, it must be fresh, with a clear and realistic shelf-life date indicated. The presence of vegetable oils in a fish oil is best tested for by examination of the sterol composition, as fish oils consist almost entirely of cholesterol and 24-methyl cholesterol (Paganuzzi, 1983). Analysis of the sterol fraction of the oil can often identify the oil, or at least indicate which oils might be present.

4.4.4 Triacylglycerol analysis of fish oils

Early studies by Litchfield (1968, 1973) showed that the exogenous long chain polyunsaturated fatty acids C22:5n–3 and C22:6n–6 followed regular distribution patterns in the TAG structures of certain aquatic species. Because of the limitations of the methods used, Takagi and Ando (1990) devised a method for stereospecific analysis of marine TAGs. Their method involved partially hydrolysing the TAGs using ethyl magnesium bromide and then resolving the MAG products, *sn*-1 and *sn*-3 fractions, by HPLC on a chiral column following derivatization. Using their method, Ando *et al.* (1992) analysed a number of fish oils, including menhaden (*Brevoortia tyrranus*) and sardine (*Sardinops melanostictus*). Their results for C20:5n–3 and C22:6n–3 are summarized in

Table 4.13 Positional distribution of EPA and DHA in menhaden and sardine oil triacyl-*sn*-glycerols (mole %)

	Menhaden oil				Sardine oil			
Fatty acid	Total	*sn*-1	*sn*-2	*sn*-3	Total	*sn*-1	*sn*-2	*sn*-3
C20:5n–3	21.20	14.43	17.40	30.87	11.66	9.01	12.20	13.72
C22:6n–3	6.44	2.67	12.86	3.60	11.53	3.08	22.07	9.09

Adapted from Ando *et al.* (1992).

Table 4.13. Both menhaden and sardine oils had C20:5n–3 and C22:6n–3 preferentially esterified at positions 3 and 2, respectively. With respect to C22:6n–3, smaller amounts were found at the other two positions with a preference for position 3. There was a clear difference between menhaden and sardine oils with respect to the relative distribution of these fatty acid in their respective triacyl-*sn*-glycerols. Based on previous research, Ando *et al.* (1992) showed the positional distribution of C20:5n–3 and C22:6n–3 was influenced by the proportion of C20:1 and C22:1 in the total triacyl-*sn*-glycerols (Brockerhoff *et al.*, 1968; Litchfield *et al.*, 1968, 1973). These results were fairly consistent with earlier reports by Brockerhoff *et al.* (1968) of a general tendency for fish oil triacyl-*sn*-glycerols to attract saturated and monounsaturated fatty acids at position *sn*-1; polyunsaturated and short fatty acids at the *sn*-2 position; and long chain fatty acids at position *sn*-3. A subsequent study on fish oil by Ando *et al.* (1996) confirmed the preference of C20:5n–3 for the *sn*–3 position and C22:6n–3 for the *sn*-2 position in sardine oil TAGs. Methods are continually being developed, modified and improved to more accurately determine the positional distribution of fatty acids in fish oil TAGs using a combination of chemical and enzymatic methods together with HPLC and ^{13}C NMR (Ando *et al.*, 2000; Amate *et al.*, 1999; Myher *et al.*, 1996; Sacchi *et al.*, 1993; Sawada *et al.*, 1993). These improvements will allow greater efficiency and accuracy in analyzing triacyl-*sn*-glycerol molecular species in fish oils to more readily establish their authenticity.

References

Ackman, R.G. and Cunnane, S.C. (1992) Long-chain polyunsaturated fatty acids: sources, biochemistry, and nutritional/clinical applications. *Adv. Appl. Lipid Res.*, **1**, 161–215.

Ackman, R.G., Hooper, S.N. and Hooper, D.L. (1974) Linolenic acid artifacts from deodorizing oils. *J. Am. Oil Chem. Soc.*, **51**, 42–49.

Amate, L., Ramirez, M. and Gil, A. (1999) Positional analysis of triglycerides and phospholipids rich in long chain polyunsaturated fatty acids. *Lipids*, **34**, 865–871.

Ando, Y., Nishimura, K., Aoyanagi, N. and Takagi, T. (1992) Stereospecific analysis of fish oil triacyl-*sn*-glycerols. *J. Am. Oil Chem. Soc.*, **69**, 417–424.

Ando, Y., Ota, T., Matsuhira, Y. and Yazawa, K. (1996) Stereospecific analyses of triacyl-*sn*-glycerols in docosahexaenoic acid rich fish oils. *J. Am. Oil Chem. Soc.*, **73**, 483–487.

Ando, Y., Satake, M. and Takahashi, Y. (2000) Reinvestigation of positional distribution of fatty acids in docosahexaenoic acid-rich fish oil triacyl-*sn*-glycerols. *J. Am. Oil Chem. Soc.*, **35**, 579–582.

AOCS (American Oil Chemists' Society) (1997) *Physical Characteristics of Oils, Fats and Waxes.* American Oil Chemists' Society.

Bandarra, M.N., Batista, I., Nunes, M.L., Empis, J.M. and Christie, W.W. (1997) Seasonal changes in lipids composition of sardine (*Sardina pilchardus*). *J. Food Sci.*, **62**, 40–42.

Beaubaire, N.A. and Simon, J.E. (1987) Production potential of *Borago officinalis* L. *Acta Horticulturae*, **208**, 101–103.

Belch, J.J.F., Ansell, D., Madhock, R., O'Dowd, A. and Sturrock, R.D. (1988) Effects of altering essential dietary fatty acids on requirements for non-steroidal anti-inflammatory drugs in patients with rheumatoid arthritis: a double blind placebo controlled study. *Ann. Rheum. Dis.*, **47**, 404–405.

Bergana, M.M. and Lee, T.W. (1996) Structure of long-chain polyunsaturated triacylglycerols by high resolution ^{13}C nuclear magnetic resonance. *J. Am. Oil Chem. Soc.*, **73**, 551–556.

Bimbo, A.P. (1989) Fish oils: past and present food uses. *J. Am. Oil Chem. Soc.*, **66**, 1717–1726.

Botazzi, F., Izzo, R. and Lotti, G. (1985) Influenza del riscaldamento sulla composizione dell' olio di *Oenothera biennis* L. *Agrochimica*, **29**, 331–340.

Brockerhoff, H., Hoyle, R.R., Hwang, P.C. and Litchfield, C. (1968) Positional distribution of fatty acids in depot triglycerides of aquatic animals. *Lipids*, **3**, 24.

Brossard, N., Croset, M., Pachiaudi, C., Riou, J.P., Tayot, J.L. and Lagarde, M. (1996) Retroconversion and metabolism of [^{13}C]22:6n–3 in humans and rats after intake of a single dose of [^{13}C]22:6n–3-triacylglycerols. *Am. J. Clin. Nutr.*, **64**, 577–586.

Brush, M.G., Watson, S.J., Horrobin, D.F. and Manku, M.S. (1984) Abnormal essential fatty acid levels in plasma of women with premenstrual syndrome. *Am. J. Obstet. Gynecol.*, **150**, 363–366.

Carrera, E., Garcia, T., Cespedes, A., Gonzalez, I., Fernandez, A., Hernandez, P.E. and Martin, R. (1999) Salmon and trout analysis by PCR–RLFP for identity authentication. *J. Food Sci.*, **64**, 410–413.

Carter, J.P. (1988) Gamma-linolenic acid as a nutrient. *Food Technol.*, **42**(6), 72–82.

Christie, W.W., Brechany, E.Y., Johnson, S.B. and Holman, R.T. (1986) A comparison of pyrrolidine and piclinyl ester derivatives for the identification of fatty acids in natural samples by gas chromatography–mass spectrometry. *Lipids*, **21**, 657–661.

Clough, P. (2001) Sources and production of specialty oils containing GLA and stearidonic acid. *Lipid Technol.*, **13**, 9–12.

Court, W.A., Hendel, J.G. and Pocs, R. (1993) Determination of the fatty acids and oil content of evening primrose (*Oenothera biennis* L.). *Food Res. Int.*, **26**, 181–186.

Cunnane, S.C. (1995) Metabolism and function of α-linolenic acid in humans, in *Flaxseed in Human Nutrition* (eds. S.C. Cunnane and L. Thompson), AOCS Press, Champaign, Illinois, USA, pp. 99–127.

Devinat, G.L., Grandgirard, A., Septier, Ch. and Prevost, J. (1980) Isomerisation de lacide linolenique durant la desodorisation des huiles de colza et da soja. *Rev. Franc. Corp Gras.*, **27**, 283–290.

De Vries, B. (1963) Quantitative separations of higher fatty acid methyl esters by adsorption chromatography on silica impregnated with silver nitrate. *J. Am. Oil Chem. Soc.*, **40**, 184–186.

Emken, E.A. (1995) Influence of linoleic acid on conversion of linolenic acid to omega-3 fatty acids in humans, in *Proceedings from the Scientific Conference on Omega-3 Fatty Acids in Nutrition, Vascular Biology, and Medicine.* American Heart Association, Dallas, Texas, USA, pp. 9–18.

Fantoni, C.M., Cuccio, A.P. and Barrera-Arellano, D. (1996) Brazilian encapsulated fish oils: Oxidative stability and fatty acid composition. *J. Am. Oil Chem. Soc.*, **73**, 251–253.

Gamez-Meza, N., Higuera-Ciapara, I., Calderon de la Barca, A.M., Vazquez-Moreno, L., Noriega-Rodriguez, J. and Angulo-Guerrero, O. (1999) Seasonal variation in the fatty acid composition and quality of sardine oil from *Sardinops sagax caeruleus* of the Gulf of California. *Lipids*, **34**, 639–642.

Gibson, R.A. and Rassias, G. (1990) In *Omega-6 Essential Fatty Acids: Pathophysiology and Roles in Clinical Medicine* (ed. D. Horrobin) Alan R. Liss, New York, pp. 283–293.

Gibson, R.A., Lines, D.R. and Neumann, M.A. (1992) Gamma linolenic acid (GLA) content of encapsulated evening primrose oil products. *Lipids*, **27**, 82–84.

Goffman, F.D. and Galletti, S. (2001) Gamma-linolenic acid and tocopherol contents in the seed oil of 47 accessions from several *Ribes* species. *J. Agric. Food Chem.*, **49**, 349–354.

Houwelingen, A.C. and Hornstra, G. (1994) Trans fatty acids in early human development. *World Rev. Nutr. Diet.*, **75**, 175–178.

Jorgensen, M.H., Hernell, O., Lund, P., Holmer, G. and Michaelsen, K.F. (1996) Visual acuity and erythrocyte docosahexaenoic acid status in breast-fed and formular-fed term infants during the first four months of life. *Lipids*, **31**, 99–105.

Kochar, S.P. (1983) Influence of processing on sterols in edible oils. *Prog. Lipid Res.*, **22**, 161–188.

Krzynowek, J., Uljua, D.S., Panuzio, L.J. and Maney, R.S. (1992) Factors affecting fat, cholesterol, and omega-3 fatty acids in Maine sardines. *J. Food Sci.*, **57**, 63–65.

Laakso, P. (1997) Characterization of α- and γ-linolenic acid oils by reversed-phase high-performance liquid chromatography-atmospheric pressure chemical ionization mass spectrometry. *J. Am. Oil Chem. Soc.*, **74**, 1291–1300.

Lawson, L.D. and Hughes, B.G. (1988) Triacylglycerol structure of plant and fungal oils. *Lipids*, **23**, 313–317.

Lee, K.T. and Foglia, T.A. (2001) Fractionation of menhaden oil and partially hydrogenated menhaden oil: Characterization of triacylglycerol fractions. *J. Am. Oil Chem. Soc.*, **78**, 297–303.

Litchfield, C. (1968) Predicting the positional distribution of docosahexaenoic and docosapentaenoic acids in aquatic animal triglycerides. *Lipids*, **3**, 417–419.

Litchfield, C. (1973) Taxonomic patterns in the triglyceride structure of natural fats. *Fette Seif. Anstrichm.*, **75**, 223–231.

Lluch-Belda, D., Megallon, J.F. and Schartlose, R.A. (1986) Large fluctuations in the sardine fishery in the Gulf of California: Possible causes. *Calif. Coop. Oceanic Fish. Invest. Rep.*, **27**, 136–140.

Lotti, G., Izzo, R. and Marchini, F. (1980) La composizione acidica delloliodi semi di *Oenothera biennis* L. durrante la maturzione. *Agrochemica*, **24**, 274–285.

Manku, M.S., Horrobin, D.F., Morse, N.L., Wright, S. and Burton, J.L. (1984) Essential fatty acids in the plasma phospholipids of patients with atopic eczema. *Br. J. Dermatol.*, **110**, 643–648.

Mannino, S. and Amelotti, G. (1975) Sterol composition of thirty vegetable oils determined by GLC using two phases of different polarity. *Riv. Ital. Sost. Grasse*, **52**, 79–83.

Mansel, R.E., Pye, J.K. and Hughes, L.E. (1990) In *Omega-6 Essential Fatty Acids: Pathophysiology and Roles in Clinical Medicine* (ed. D. Horrobin) Alan R. Liss, New York, pp. 557–566.

Marin, R.E. and Flick, G.J.A. (1990) History of Seafood Industry, in *The Seafood Industry* (eds. R.E. Marin and G.J.A. Flick) Van Nostrand Reinhold, New York, pp. 8–10.

Mounts, T.L. (1994) *Codex* fats and oils panel meets in London. *Inform*, **1**, 96.

Muderhwa, J.M., Dhuique-Mayer, C., Pina, M., Galzy, P., Grignac, P. and Graille, J. (1987) Repartition interne/externe ders acides gras des triglycerides de quelques huiles gamma linoleniques. *Oleagineux*, **42**, 207–211.

Myher, J.J., Kuksis, A., Gehr, K., Park, P.W. and DiersenSchade, D.A. (1996) Stereospecific analysis of triacylglycerols rich in long-chain polyunsaturated fatty acids. *Lipids*, **31**, 207–215.

Ockerman, P.A., Backrack, I., Glans, S. and Rassner, S. (1986) Evening primrose oil as a treatment of premenstrual syndrome. *Recent Adv. Clin. Nutr.*, **2**, 404–405.

Paganuzzi, V. (1983) Analysis of the sterol fraction of olive oil used for covering canned fish. *Riv. Ital. Sost. Grasse*, **60**, 116–124.

Ratnayake, W.M.N., Olsson, B., Matthews, D. and Ackman, R.G. (1988) Preparation of omega-3 PUFA concentrates from fish oils via urea concentration. *Fat Sci. Technol.*, **90**, 381–386.

Rehbein, H., Kress, G. and Schmidt, T. (1999) Application of PCR–SSCP to species identification of fishery products. *J. Sci. Food Agric.*, **74**, 35–41.

Reina, R.J., White, K.D. and Jahngen, E.G.E. (1997) Validated method for quantitation and identification of 4,4-desmethylsterols and triterpene diols in plant oils by thin-layer chromatography–high resolution gas chromatography–mass spectrometry. *J. AOAC Intern.*, **80**(6), 1272–1280.

Rossell, J.B. (1991) Purity criteria in edible oils and fats. *Fat Sci. Technol.*, **93**, 526–531.

Sacchi, R., Medina, I., Aubourg, S.P., Guidicianni, I., Paolilli, L. and Addeo, F. (1993) Quantitative high resolution ^{13}C NMR analysis of lipids extracted from the white muscle of Atlantic tuna (*Thunnus alalumga*). *J. Agric. Food Chem.*, **41**, 1247–1253.

Sawada, T., Takahashi, K. and Hatano, M. (1993) Triglyceride composition of tuna and bonito orbital fats. *Nippon Suisan Gakkaishi*, **59**, 285–290.

Sensidoni, A., Bortolussi, G., Orlando, C. and Fantozzi, P. (1996) Borage oil (*Bortago officinalis* L.); an important source of gamma linolenic acid. II. Tocopherols and chlorophyll content and sensorial analysis of borage oils extracted by different techniques and blended with extra virgin olive oil. *Ind. Aliment.*, **35**, 664–669.

Senanayake, S.P.J.N. and Shahidi, F. (2000) Lipid components of borage (*Borago officinalis* L.). *J. Am. Oil Chem. Soc.*, **77**, 55–61.

Shukla, V.K.S. and Perkins, E. (1991) The presence of oxidative polymeric materials in encapsulated fish oil. *J. Am. Oil Chem. Soc.*, **26**, 23–26.

Singer, P., Moritz, V., Wirth, M., Berger, I. and Forster, D. (1990) Blood pressure and serum lipids from SHR after diets supplemented with evening primrose, sunflower seed or fish oil. *Prostaglandins Leukotrienes Essential Fatty Acids*, **40**, 17–20.

Stansby, M.E. (1981) Reliability of fatty acid values purporting to represent composition of oil from different species of fish. *J. Am. Oil Chem. Soc.*, **58**, 13–16.

Tagaki, T. and Ando, Y. (1990) Enantiomer separations of mixtures of monoacyl derivatives by HPLC on a chiral column. *Lipids*, **21**, 413.

Traitler, H., Winter, H., Richli, U. and Ingenbleek, Y. (1984) Characterization of γ-linolenic acid in *Ribes* seeds. *Lipids*, **19**, 923–928.

Trautwein, E.A. (2001) n–3 Fatty acids—physiological and technical aspects for their use in food. *Eur. J. Lipid Sci. Technol.*, **103**, 45–55.

Uzzan, A., Helme, J.P. and Klein, J.M. (1992) Fatty acid composition and quality data for oils rich in GLA. *Rev. Franc. Corps Gras*, **39**(11/12), 339–343.

Wolff, R.L. (1992) *trans*-Polyunsaturated fatty acids in French edible rapeseed and soybean oils. *J. Am. Oil Chem. Soc.*, **69**, 106–110.

Wolff, R.L. (1993a) Occurrence of artificial *trans* polyunsaturated fatty acids in refined (deodorized) walnut oils. *Sci. Aliment.*, **13**, 155–163.

Wolff, R.L. (1993b) Further studies on artificial geometrical isomers of α-linolenic acid in edible linolenic acid containing oils. *J. Am. Oil Chem. Soc.*, **70**, 219–224.

Wolff, R.L. and Sebedio, J.L. (1991) Geometrical isomers of linolenic acid in low-calory spreads marketed in France. *J. Am. Oil Chem. Soc.*, **68**, 719–725.

Wolff, R.L. and Sebedio, J.L. (1994) Characterization of γ-linolenic acid geometrical isomers in borage oil subjected to heat treatments (deodorization). *J. Am. Oil Chem. Soc.*, **71**, 117–126.

Wretensjo, I. and Svensson, L. (1990) Gas chromatographic-mass spectrometric identification of the fatty acids in borage oil using the picolinyl ester derivatives. *J. Chromatogr.*, **521**, 80–97.

5 Milk fat and other animal fats

Michael Jee

5.1 Introduction

Fats of animal origin were the first to be used for edible purposes. Although in recent years the volume of these fats produced has stayed approximately constant (or increased only slightly), the demand for them for edible purposes has in most cases decreased—due at least in part to perceived or potential health problems. This is because all fats of animal origin contain cholesterol as the major sterol, a constituent that is normally considered to be of negative value in the diet and because, with the exception of most marine fats, they are high in saturated fatty acids, which is also seen as a negative attribute.

Animal fats fall naturally into three main groups: marine, milk and carcass fats.

(a) Marine fats can be from any variety of marine animal. They contain cholesterol, and the fatty acids usually contain significant levels of highly unsaturated omega-3 fatty acids, such as eicosahexaenoic and docosapentaenoic acids. Their authenticity is dealt with in Chapter 4 of this volume.
(b) Milk fats in western society are usually taken to mean cow milk fat. Milk fats from other animals, such as goats, sheep, camel and buffalo, are becoming recognized, but still represent a relatively small part of the market. They contain cholesterol and a wide range of fatty acids, a large proportion of which are saturated.
(c) Carcass fats can originate from any animal. The major sources are beef cattle (beef tallow), sheep (mutton tallow), goats (goat tallow) and pigs (lard). Certain other grades of fat from these sources, such as greases or neatsfoot oil, are produced, but these are not usually edible fats. Edible fats from other animals, such as goose and chicken, are available, but they have a specialized market, and only relatively small amounts are produced and sold. They all contain cholesterol, and are mostly high in saturated fatty acids, though the levels vary.

5.2 Checking for the absence of animal fats

5.2.1 Requirements

Animal fats are the only group of fats where their absence, or the absence of one type (pork), can be a desirable attribute, as opposed to the more normal requirements for authenticity, where it is the presence (usually at 100% as far as can be determined) of a fat that is desirable. These requirements are usually for ethical or religious reasons. Many do not wish to consume animal products. In some cases this requirement concerns all products of animal origin, including milk and eggs. In other cases it is merely necessary to abstain from consumption of products arising from the death of an animal. In the latter case only carcass fats and marine oils are a concern.

The necessity for non-consumption of fat from a particular source is usually religious, most commonly from Jews and Moslems. The religious laws forbid the consumption and use of pork products. There are also requirements concerning shellfish, but shellfish oil is not, and would be unlikely to ever be, a commercial product.

5.2.2 Determining the absence of any animal (including marine) fats

It has previously been stated that one of the reasons for the decline in animal fat consumption is the presence of cholesterol. It is this aspect that can be used to identify the presence of animal fats in a fat or product. There are some older methods of examination of sterols, including cholesterol, and these are still listed and may still be used. They usually involve determination of the melting points of the sterol acetates after extraction from the fat using a digitonin column (AOAC, 1995a) or use thin layer chromatography (TLC) to detect the acetates (Mathew and Kamath, 1978). However this method has been largely superseded by analysis of sterols (usually silylated) using gas chromatography (GC). The older method of extraction using a digitonin column (LaCroix, 1970; Nakazawa, 1981) can still be used. However, more usually, when an oil is being examined, the sample is saponified, the unsaponifiable matter usually extracted with solvent, and the sterol fraction separated by TLC using a suitable eluent such as diethyl ether/hexane (35:65) or benzene/acetone (95:5). The sterol band may be visualized by spraying with dichlorofluorescein or some similar spray reagent, scraped off the plate and extracted. It is often then silylated and run on a slightly polar stationary phase such as Se-54, CpSil8, OV-5 or DB-5. Both digitonin and saponification methods give similar results for free cholesterol (Homberg and Bielefeld, 1987), but, if it is the absence of total cholesterol that is required, then a saponification method should be preferred. For quantitation purposes some methods (AOAC, 1995b; Thorpe, 1970) add an internal standard (5α-cholestane) after the TLC stage. Addition at this stage rather than before is because this compound does not separate with the sterol band in the TLC, but addition here does introduce a possible source of error. If the TLC purification step is omitted, then the α-cholestane internal standard can be added initially to the oil before saponification (Youk *et al.*,1999). This does introduce the possiblity of interference from co-eluting non-sterol components, but no interferences have been reported for cholesterol, though some terpene alcohols do interfere with other sterols. An alternative method (EEC, 1991) uses an internal standard which does run with the sterol band in the TLC separation (dihydrocholesterol). Dihydrocholesterol does elute on the GC very close to cholesterol, and so, if this internal standard is used, a column and conditions giving good resolution are required to ensure that the two peaks separate well. A third internal standard used in sterol analysis, which does separate on TLC with the sterol, is betulin (EC, 1991). This however elutes on the GC much later than cholesterol, and is therefore normally considered unsuitable for this analysis, although its use in the analysis of lard and butter has been reported (Homberg and Bielefeld, 1987). The saponification step does ensure that all the cholesterol, including esterified cholesterol, is observed.

A method of separation of the sterol fraction using a chromatographic module containing a silica column that can be coupled to a GC has also been described (Ballesteros *et al.*, 1995). This has been used for cholesterol determination in addition to determination of other sterols, though cholesterol esters would elute separately from cholesterol. These esters make up about 10% of the cholesterol of milk fat (Gunstone *et al.*, 1994). It is possible to analyse sterols without any separation of other components, although again this is only satisfactory for free cholesterol, cholesterol esters eluting separately. This has been carried out by direct on-column injection of the fat onto a non-polar column fitted with a pre-column (Grob, 1984). The triglyceride remains on the pre-column. The pre-column has to be regularly replaced, and so the method is not suitable for continual unattended analysis of large numbers of samples. Alternatively no pre-column may be used (Chung and Lucy, 1996), though this may reduce the life of the column. This method has been reported to have been used for animal and egg fats and there do not appear to be interfering compounds present that might cause problems due to co-elution with cholesterol or with the α-cholestane internal standard.

5.2.3 *Interpretation of the results of cholesterol determinations*

The result of the analysis will determine the level of cholesterol present, if any. It has long been generally accepted (Guyot, 1969; Karleskind, 1969) that animal fats are the only fats in which cholesterol is the major sterol present. Milk fat also contains traces of lanosterol and dihydrolanosterol. Therefore the absence of cholesterol is reasonable evidence that animal fats are absent. Problems arise in interpretation in mixtures of fats, when cholesterol is present at a low level as a proportion of the total sterols. Some vegetable fats have been reported (Kanematsu *et al.*, 1973; Seher, 1987) to contain low levels of cholesterol (<3% of total sterols in oils such coconut, palm kernel and rapeseed), and slightly higher levels of up to 7% have been reported for palm oil (Homberg, 1991). Thus, if an oil contains cholesterol at this level, it is necessary to weigh this against the composition expected, possibly after fatty acid analysis, and reach an educated decision based on experience. There is only one vegetable oil reported to contain high levels of cholesterol (Kiosseoglou and Boskou, 1987). They claimed that in tomato seed oil cholesterol comprised up to 41% of esterified sterols. Other workers reported cholesterol as 7–27% of total sterols (Tiscornia *et al.*, 1976), but even if 41% is normal in this oil, the oil is not likely to be an article of commerce, or present at any significant level in any oil blend. It is possible that results would be affected on a product containing a high level of tomatoes and low levels of fat, but this would be a very unusual situation.

Various processes have been reported that claim to reduce the cholesterol level of animal fats significantly using supercritical carbon dioxide extraction (Camin *et al.*, 2000), β-cyclodextrin complexation (Gow and Li, 1995; Yen and Chen, 2000), ethanol extraction (Youk *et al.*, 1995), reaction with succinic anhydride (Wrezel *et al.*, 1992) or by molecular distillation (Gu *et al.*, 1994; Lanzani *et al.*, 1994), while it has been suggested that genetic modification might reduce levels (Gomer and Kotulak, 1991). Removal of over 75% seems to be routinely claimed, and this might be expected to give rise to problems in determining the absence of animal fats in a fat, oil or product, as some vegetable products do apparently contain low levels of cholesterol, or a sterol that is apparently cholesterol (Kanematsu *et al.*, 1973). It is certainly true that fats which had been treated in one of these ways, when blended in as only a part of the fat, could possibly not be detected. However, the treatment of fats by these processes would make the products much more expensive than the original animal products, and it is doubtful if the use of these modified fats as a small part of a blend could be justified economically. Therefore it is extremely unlikely that the problem would arise, and it is a reasonable presumption that, if significant levels of cholesterol are absent in a fat, then animal fat is absent.

5.2.4 Absence of animal fats in oleochemicals

Sometimes the question asked is whether animal fat is absent from products such as cosmetics or from some other oleochemical product prepared from fats. This can include products such as emulsifiers. Where the product contains original triglyceride then this portion can be considered as described above, and the absence of cholesterol can be considered as good evidence that animal fats are absent. This, however, does not apply to triglycerides formed after saponification of fatty acids, followed by fractionation and recombination with glycerol. So-called 'fractionated coconut oil' is manufactured by this process, and any similar product made from fatty acids from an animal source would not show the presence of any cholesterol. Similarly other oleochemicals formed from fatty acids derived from animal fats would also not contain cholesterol. In these cases it is often impossible to detect the presence of animal fat by testing for cholesterol. The only possibility in some cases might be to look at

Table 5.1 Fatty acid composition of glycerol monostearate from different batches originating from one of three different oil sources

Fatty acid	A	B	C	D	E	F	G	H	I	J
C10:0	0.1	<0.1	<0.1	<0.1	<0.1	0.1	0.1	<0.1	0.1	0.1
C12:0	0.3	0.2	0.1	0.1	0.1	0.3	0.3	0.1	0.1	0.1
C13:0	<0.1	ND	ND	ND	ND	<0.1	<0.1	<0.1	<0.1	<0.1
C14:0b	<0.1	ND	ND	ND	ND	<0.1	0.1	<0.1	<0.1	<0.1
C14:0	3.7	3.2	1.0	1.1	0.9	3.8	3.8	3.7	1.4	1.4
C14:0I	0.2	ND	ND	ND	ND	0.2	0.2	0.2	<0.1	<0.1
C15:0b	0.2	ND	ND	ND	ND	0.2	0.3	0.2	<0.1	<0.1
C15:0	0.4	0.4	0.1	0.1	0.1	0.5	0.5	0.5	0.1	0.1
C16:0b	0.1	0.2	ND	ND	ND	0.1	0.2	0.2	0.1	0.1
C16:0	29.7	30.5	41.9	43.7	42.8	29.5	28.2	27.7	26.4	26.4
C16:0I	0.3	0.3	ND	ND	ND	0.3	0.4	0.3	<0.1	<0.1
C17:0b	0.5	0.6	ND	ND	ND	0.6	0.7	0.6	0.1	0.1
C17:0	0.1	0.1	ND	ND	ND	0.1	0.1	0.1	ND	ND
C17:0I	1.3	1.4	0.1	0.1	0.1	1.5	1.6	1.9	0.7	0.7
C18:0b	0.1	0.1	ND	ND	ND	0.2	0.2	0.2	0.1	0.1
C18:0	60.7	60.9	55.9	54.2	55.3	60.5	61.6	62.4	68.1	68.0
C18:1	0.1	<0.1	0.1	0.1	0.1	0.1	0.1	0.1	0.1	0.1
C19:0	0.3	0.3	<0.1	ND	ND	0.3	0.3	0.3	0.2	0.2
Unid.	0.1	ND	ND	ND	ND	0.1	0.1	0.1	<0.1	<0.1
C18:2?	<0.1	0.1	ND	ND	ND	<0.1	<0.1	<0.1	<0.1	ND
C20:0	1.2	1.0	0.6	0.5	0.6	1.1	0.9	0.8	2.3	2.3
C21:0	<0.1	ND	ND	ND	ND	<0.1	<0.1	<0.1	<0.1	<0.1
C22:0	0.3	0.1	0.1	<0.1	0.1	0.2	0.2	0.1	0.3	0.3

I, Iso; B, Branched; Unid, unidentified; ND, no data.

the fatty acid composition. Where the fatty acid used to produce the product is derived entirely, or largely, from animal sources, then the presence of minor odd-numbered or branched fatty acids can indicate the presence of animal fat. Glycerol monostearate (GMS) is manufactured from fully hydrogenated oils, which can originate from vegetable or animal sources. Table 5.1 shows the fatty acid composition of different batches of GMS which were being investigated in order to determine whether a correlation could be found of the composition with the deterioration of the product in which they were used. The batches fell into three groups. Some samples (C, D and E), where the only significant odd-numbered fatty acids were C15:0 and C17:0, also contained high levels of C16:0, indicating that the product was manufactured from palm oil only. The second group (A, B, F, G, H) contained low but detectable levels of C13:0, C14:0-branched, C15:0-branched, C16:0-iso, C17:0-iso, and C18:0-iso, C19:0 and C21:0 and much higher levels of C15:0 and C17:0. This is good evidence that the second group is manufactured from animal fat. The levels of C14:0 would indicate that tallow was probably the source. The third group (H and I) had much lower levels of odd-numbered and branched fatty acids than group 2; the levels of C18:0 and C16:0 were similar to group 2. It is possible that these were manufactured from a blend of animal and vegetable oils, but this is uncertain.

5.2.5 Absence of pork fat in oil

Naturally, if no animal fat can be detected after cholesterol analysis, then it can reasonably be concluded that pork fat is not present. If cholesterol is present, either as a low percentage of the total sterols or as a major or only component of the sterols, then other approaches have to be employed. If the product is likely to contain any DNA material, then DNA analysis would probably be the best approach (Montiel-Sosa *et al.*, 2000). This might include rendered fat if the fat had not been highly refined, though no work appears to have been done to determine whether this would be possible.

It is claimed by some laboratories that they can identify the absence of pork fat in, for example, beef tallow, by the determination of the simple fatty acid composition with a rough comparison of fatty acid compositions. Certainly the accepted ranges for lard and edible tallows (including premier jus) given in *Codex Alimentarius* are different (Table 5.2.), and pure fats can often be differentiated in this way, but the natural variability of the product ensures that this is not certain when lower levels of pork fat (<50%) are present. In the

Table 5.2 Acceptable ranges of fatty acid composition for pork fat (lard) and beef fat (beef tallow and premier jus) given in *Codex Alimentarius* (*Codex Alimentarius* Commission, 1993a,b,c)

Fatty acid	Lard	Premier jus and beef tallow
C<14	<0.5	<2.5
C14:0	0.5–2.5	1.4–7.8
C14:1	<0.2	<0.3
C15:0	<0.1	0.5–1.5
C15:0 ISO + ANTE-ISO	<0.1	<1.5
C16:0	20–32	17–37
C16:1	1.7–5.0	0.7–8.8
C16:2	—	<1.0
C16:0 ISO	<0.1	<0.5
C17:0	<0.5	0.5–2.0
C17:1	<0.5	<1.0
C17:0 ISO + ANTE-ISO	—	<1.5
C18:0	5.0–24	6.0–40
C18:1	35–62	26–50
C18:2	3.0–16	0.5–5.0
C18:3	<1.5	<2.5
C20:0	<1.0	<0.5
C20:1	<1.0	<0.5
C20:2	<1.0	—
C20:4	<1.0	<0.5
C22:0	<0.1	—

case of both major and minor fatty acids present in animal fats, the levels have been shown to be influenced considerably by the diet of the animal, and even branched chain fatty acids can be increased by the feeding of ruminant fat to pigs (Bastijns, 1968, 1970). The matter can be further complicated by the possibility of the presence of a vegetable oil, or several vegetable oils in the oil.

It has been suggested that lipolysis of the fat with pancreatic lipase followed by GC of the resulting 2-monoglyceride can detect the presence of lard in beef products (Dabash *et al.*, 1979; Javidipour *et al.*, 1999). It is necessary to determine the enrichment factor (percentage of total fatty acids in 2-monoglyceride per percentage in whole triglyceride) for palmitic acid, enrichment value of unsaturation ratio, and ratio of C16/C18 fatty acids in the monoglycerides. These values are: <0.8 for beef; >0.8 for pork; <1.4 if lard is present; and <0.7 for beef, >4 for lard, respectively. Regression analysis showed that levels down to 10% lard could be detected (Verbeke and de Brabander, 1979) though this was only using test mixes and did not include any vegetable fat, though similar work on lard in buffalo tallow was also reported (Youssef and Rashwan, 1989).

Two other methods also based on differences in the levels of saturated fatty acids in the 2-position have also been proposed for detecting pork fat in beef and sheep products. Ozonolysis of the fat, followed by separation of the ozonides and the unaffected saturated triglycerides on a reversed phase column, detects and differentiates unsymmetrical triglycerides, in which the unsaturated fatty acid is at the 1 or 3 positions (saturated–saturated–unsaturated or SSU), and symmetrical triglycerides, where the unsaturated fatty acid is at the 2 position (saturated–unsaturated–saturated or SUS). For manufactured mixtures, levels down to 3% pork fat can be detected. Studies of vegetable fats carried out to determine the position of the oleic acid (1+3, or 4 positions) used another procedure than ozonolysis for differentiating the two classes of triglycerides (Podlaha and Toregard, 1984). Here halogen-addition to the double bonds was used. This is probably easier to carry out than ozonolysis, gives a less complicated high performance liquid chromatography (HPLC) trace, and appears to give reasonable separation. Therefore it would be expected to work in the detection of pork fat. The procedure of separation of symmetrical and unsymmetrical triglycerides could now undoubtably be carried out using silver ion HPLC, but I do not believe that this has yet been tried, or, at least, been reported.

The last suggested method of detection (Saeed *et al.*, 1986) involves detection of eicosa-11,14-dienoic acid, which is present in pork fat but not in tallow or mutton fat. It is claimed that down to 1% pork in beef can be detected, but the results would certainly be affected by the presence of significant levels of this acid in some vegetable fats in which it is present at low levels (Gunstone *et al.*, 1994). Therefore, although the method would be useful in confirming the absence of pork fat, if vegetable fats were also present the method might give false positives. If this method was to be used, and a positive result for pork fat was found, then sterol analysis would also have to be carried out to ensure the absence of vegetable fats.

The problem of the presence of pork fat derivatives in oleochemicals and products is similar to that already described with respect to animal oleochemicals in general. A summary of the status of food emulsifiers with respect to Jewish food laws is given by Hodd (1996). The author, however, presumably obtained his information from the manufacturers and not by testing the products.

It may therefore be stated that there are a number of methods that might be used to confirm the absence of pork fat, but that none of them can be considered to give a definitive answer. When the question is asked as to the absence of oleochemicals derived from pork fat, then there is no way of confirming their absence.

5.3 Authentication of milk fats

5.3.1 Bovine milk fat

The fat from milk from all sources for which analyses are available contains short chain fatty acids, presumably because these are more easily absorbed by the young animal. Some other oils such as palm kernel and coconut species contain acids of carbon length C6–C14, but only milk fats contain butyric acid (C4:0). In bovine milk fat the percentage of each short chain, even-numbered fatty acid present in the milk increases approximately as the carbon number of the acid increases. This is unlike oils such as palm kernel or coconut, where, as the carbon number increases, there is a peak at lauric acid (C12:0). In these oils lauric acid is usually around 45% of the total fatty acids present. Because of this, methods for estimation of milk fat content have usually relied in some way on the amount of butyric acid. This was initially by means of the Reichert value, where the water-soluble/steam-distillable acids are determined, and later by gas chromatographic determination either as the free acid, methyl ester or butyl ester (Hadorn and Zurcher, 1972; Iverson and Sheppard, 1986). In these methods the percentage of butyric acid present in milk fat is usually taken as about 3.6%. However, as the value can vary over a range, this determination is of no use in determination of the complete authenticity of the milk fat, and other techniques have to be used for this purpose. Similar factors affect the use of the triglyceride group C34, sometimes used roughly to determine percentage of milk fat in mixtures (Schneller and Wullschleger, 1992), although here interference is also found when lauric fats (coconut and palm kernel) are present.

5.3.1.1 Vegetable fats in milk fat

In testing for authenticity of bovine milk fat (Luf, 1988a; Younes and Soliman, 1987) the first test should be for the presence of plant sterols, this indicating the presence or absence of vegetable oils. This can be carried out by similar procedures to those used to determine cholesterol. The oil is saponified, solvent extracted, TLC used to separate the sterol fraction, and GC carried out on a non-polar capillary column, either as unreacted sterols or after silylation, using an internal standard for quantitation (AOAC, 1995b; EEC, 1991). Unlike where the sterol of interest is cholesterol, the TLC stage is essential, as other components could interfere with plant sterol peaks. This can be seen in some oils if the 'terpene alcohol' band eluting above the sterols on the TLC plate is run on the GC column used to separate the sterol band. Overlap of components from the two fractions can, on occasion, be seen to occur.

If plant sterols are found then it is reasonable to conclude the presence of adulteration (Hallabo and El-Nikeety, 1987; Luf, 1988a; Younes and Soliman, 1987). It has been claimed (Parodi, 1973a) that traces of plant sterols were found in milk fat, but other workers appear not to have confirmed their presence, and lanosterol, which can be present in trace levels in milk fat, elutes close to β-sitosterol (Homberg, 1991; Ulberth and Buchgraber, 2000). It is reasonable to conclude therefore that the components identified were either artefacts (possibly arising from poor TLC separation or accidental contamination) or misidentified other components. If the vegetable oil present can be identified from sterol and fatty acid analysis, then estimates of the level of adulteration can be made. These estimates will usually be very rough, however, as there is a large variation in levels of sterols present in different samples of the same vegetable oil (*Codex Alimentarius* Commission, 1997; Gunstone *et al.*, 1994).

If no plant sterols are found in the sterol fraction, then untreated vegetable oils can reasonably be ruled out as being present. It is possible that the treatments described earlier which reduce cholesterol in animal fats (section 5.2.2) could also do the same with plant sterols, but the costs would almost certainly be prohibitive. It is, however, possible to remove the sterols of vegetable oils chemically, and in the past this has been carried out to hide the adulteration of olive oils (Firestone, 2001). The process involves treatment of the oil with high levels of bleaching earths at high temperatures, but also results in the formation of sterene hydrocarbons, mainly stigmasta-3,5-diene, campesta-3,5-diene and stigmasta-3,5,22-triene. However, these can be detected by a combination of column chromatography and GC on a slightly polar column, possibly with GC/MS confirmation (Bonvehi *et al.*, 2001; Cert *et al.*, 1994; Firestone, 2001; Grob *et al.*, 1994; Lanzon *et al.*, 1994). There is a problem with high oleic sunflower oil, where isomerization, rather than dehydration, occurs, but this problem has also been solved (Mariani and Ventturini, 1997). It is possible that desterolized oils might be used in milk fat adulteration and, as a precaution, if there is any doubt, possibly the sterene hydrocarbon level should be investigated, even though it is likely that triglyceride analysis would also detect adulteration from the oil. Detection of cholesta-3,5-diene has also been used to detect the presence of refined beef tallow in milk fat (Mariani *et al.*, 1994).

5.3.1.2 *Fatty acid analysis of milk fat*

Fatty acid analysis of a fat is nowadays a relatively routine analytical operation. After methylation of the fat using reaction with boron trifluoride/methanol, boron trichloride/methanol, methanolic hydrogen chloride solution, diazomethane or, if free fatty acids are not present, alkaline catalysts such as sodium methoxide/methanol, the prepared methyl esters are then analysed by GC on a polar column such as CpSil 88, BPX70 or SP2340. The high polarity of the column is necessary to separate the saturated and unsaturated fatty acids fully. The fatty acid composition of a milk fat sample is thus relatively easily obtained, and was therefore one of the first techniques investigated for authentication purposes.

Milk fat does have a very characteristic fatty acid composition, and contains about 15 major fatty acids and several hundred minor fatty acids (Gunstone *et al.*, 1994; Hettinga, 1996). One might think that this would mean that authentication would be relatively easy from just the fatty acid composition. However the fatty acid composition is not just complicated, it is also very variable. Early surveys showed this variability (Hilditch and Williams, 1964; Hallabo and El-Nikeety, 1987) and some of the reasons for this variability were soon evaluated (Hilditch and Williams, 1964).

Some of the principal differences are brought about by diet. A diet of either silage or roots can cause considerable differences in composition (Hilditch and Williams, 1964). Silage increases the levels of *cis*-C18:1 and C18:0, but reduces C16:0 acid, with roots having the opposite effect. This in turn causes variations over the year, as diet often changes depending on the diet provided for the animals (Fox *et al.*, 1989; Sieber *et al.*, 1998). When particular fats are fed then the composition can change considerably. Rapeseed oil decreases C16:0 and increased *cis*-C18:1, while, if the oil was available in the rumen, then C18:0 acid also increased (DePeters *et al.*, 2001; Jenkins, 1998; Murphy *et al.*, 1995). Soyabeans are reported to give variable results when fed (Murphy *et al.*, 1995; Sol-Morales *et al.*, 2000). Fish oil decreased C18:0 and *cis*-C18:1, but increased *trans*-C18:1, conjugated linolenic acid and C18:3 (unconjugated). Surprisingly, considering the hydrogenation occurring in the rumen caused by microorganisms, it also increased C20:4 and C20:5, although less than 7% was transferred to the milk (Jones *et al.*, 2000; Offer *et al.*, 1999). Chilliard *et al.*, (2000) have summarized the various alterations that can be made to the fatty acid composition of milk using dietary factors, while Hermansen (1995) attempted to predict the fatty acid composition of a milk from the feed composition. Problems have been reported in some countries because certain dietary regimes produced milk fat compositions which were non-compliant with local regulations for milk fat composition (Polidori *et al.*, 1996). Similar problems occurred with butters produced from milk originating at high altitudes (Collomb *et al.*, 1999; Valentinis, 1970), where the difference are explained by highland pastures containing more non-leguminous plants, while lowland regions containing more leguminous plants.

Other factors are also claimed to affect the fatty acids in milk. Cattle low in copper are reported to give milk higher in conjugated linolenic acid (Sol-Morales *et al.*, 2000). This may be due to either an increased tendency to oxidation of the acid by the known pro-oxidant copper, or to some effect on the process of desaturation of acids in the cow. The breed of cattle also affects the composition. Fatty acids C4:0–C14:0 and C18:0 are higher in milk from Jersey compared with Holstein cows, C16:1 and *cis*-C18:1 are lower (Sol-Morales *et al.*, 2000), while *cis*-C18:1 is increased more by increased fat in the diet in Holstein compared with Jersey cattle. Similarly the ratio of C18:1/C18:0 is increased with increased fat in the diet in Holsteins, and unaffected in Jersey cattle treated to the same changes in diet. It has also been shown that cows bred for high fat milk gave lower percentages of *cis*-C18:2 and *cis*-C18:3 acids, and higher levels of C16:0 than similar cows bred for low fat milk.

The effect of added polyunsaturated acids to the diet on the composition of the milk is usually less than might be expected (Hagemeister *et al.*, 1991; Jones *et al.*, 2000; Offer *et al.*, 1999). This is due to the action of microorganisms in the rumen, which hydrogenate or otherwise affect a considerable proportion of these acids. This was shown by feeding of oils directly bypassing the rumen (Hagemeister *et al.*, 1991). Dietary protected oils fed to cattle enabled the incorporation of these acids into the milk fat, thus producing a product which has different, and for some purposes better, physical characteristics (Fearon *et al.*, 1994).

A summary of the ranges of major fatty acids (>1%) that have been found in bovine milk fats is given in Table 5.3. This includes values obtained from

Table 5.3 Range of major fatty acids reported in bovine milk fat

Fatty acid	Range of values reported (weight %)
C4:0	2.40–4.22
C6:0	2.24–2.92
C8:0	1.04–2.34
C10:0	2.12–4.21
C12:0	2.20–5.52
C14:0	9.01–17.52
C14:1	1.34–3.21
C15:0	1.09–1.29
C16:0	10.72–41.71
C16:1	2.29–3.93
C17:0	0.69–1.08
C18:0	2.78–18.81
C18:1 *trans* + *cis*	11.63–35.75
C18:2 *trans* + *cis*	0.73–5.52
C18:3	0.10–1.62

Values are a combination of values given in a range of publications, including those in which protected oils have been fed to the cattle. The results from experiments that would never be likely to be used in commercial practice are not included. These include protected tallow and coconut oil.

diets that might be used in normal commercial practice, but omits those never likely to be used commercially. Thus results from cattle fed protected soya and rapeseed oil are included because this type of diet can be used to produce soft margarines, but results from similar diets where the protected oil is coconut or tallow have been omitted because there is no commercial reason for this to be done, and indeed there are commercial reasons for it not to be done. Relatively high levels of other animal fats, and many vegetable fats, may be detected even with a large range of legitimate values for fatty acids (Fox *et al.*, 1989; Toppino *et al.*, 1982; Vallejo-Cordoba, 1999; Vanoni *et al.*, 1978) but, when only fatty acid composition is employed, lower levels of adulteration cannot be detected (Contarini *et al.*, 1999; Lipp, 1995; Precht, 1991) even when linear discriminant analysis is employed (Ulberth, 1994).

It has been suggested that examination of the fatty acids at the 2-position in addition to the total composition might help the authentication of milk fat possibly adulterated with tallow or other oils (Movia and Remoli, 1977; Soliman and Younes, 1986), but this has not been followed up further in the literature. One other approach would be to fractionate the milk fat and detect minor components. Farag *et al.*, (1983) used fractional crystallization from silver nitrate-saturated methanol/acetone. Urea complex formation (Iverson *et al.*, 1965; Iverson and Weik, 1967) might also be investigated. Values obtained on one sample of milk fat are given in Tables 5.4a and 5.4b. It would be necessary to develop a reproducible regime for fractionation, but it is possible that this might be useful in identifying low levels of adulteration where the possible adulterant is known.

5.3.1.3 Triglycerides of milk fat

When it was found that extreme variations of composition precluded the use of fatty acid analysis for milk fat authentication, the possibility of analysis of whole triglycerides was investigated. Triglycerides are more difficult to separate and analyse satisfactorily by GC. Because of the high molecular weight they are only volatile in their entirety at very high temperatures. Triglycerides of carbon number 54 (corresponding to a molecule containing three C18 fatty acids, as the carbon atoms from the glycerol are not by convention counted in triglyceride analysis) need temperatures of at least 370°C for complete elution, while, even at that temperature, on-column losses of higher molecular weights (C56, C58) can occur. Early separations involved taking the most stable column packing available (SE-30, OV-1) and conditioning the column by stripping off all the lower boiling material so as to leave a higher boiling fraction coated on the packing. This made the preparation of two reproducible columns very difficult to achieve. Now that purer, more stable, phases are available, this unsatisfactory procedure is no longer necessary, and in addition to reproducible packed columns, reproducible capillary columns are also available coated with phases such as HT-5, which can give satisfactory results at over 400°C. Even with

Table 5.4a Fatty acids reported to be present one sample of milk fat (C4–C7 omitted)

Saturated straight chain		Saturated branched		Saturated multi-branched	
Fatty acid	Weight %	Fatty acid	Weight %	Fatty acid	Weight %
8:0	0.69	12:0iso	0.01	16:0br3	0.01
10:0	1.88	13:0iso	Trace	17:0br3	0.01
11:0	0.12	14:0iso	0.03	18:0br3	0.16
12:0	2.96	14:0anteiso	0.23	19:0br4	0.02
13:0	0.10	15:0iso	0.14	20:0br4	0.14
14:0	11.20	16:0iso	0.20	21:0br4	0.02
15:0	1.52	17:0iso	0.36	22:0br4	0.02
16:0	27.8	17:0anteiso	0.23	23:0br4	0.01
17:0	0.71	18:0iso	0.02	24:0br4	0.10
18:0	12.1	19:0br	0.01	25:0br4	0.10
19:0	0.05	20:0br	0.01	26:0br3	0.01
20:0	0.02	21:0br	0.01	26:0br4	0.04
21:0	0.06	22:0br	0.02	27:0br4	0.04
22:0	0.04	23:0br	0.01	28:0br3	0.02
23:0	0.01	24:0br	0.02	28:0br4	0.12
24:0	0.02	25:0br	0.0004	28:0br5	0.01
25:0	0.02	26:0br	0.0004		
26:0	0.02				
27:0	0.0004				
28:0	0.0004				

Values obtained from Iverson *et al.*, 1965.
br, iso and/or anteiso; number after br is number of methyl branches.

Table 5.4b Minor fatty acids reported to be present in milk fat

Monoene fatty acids		Diene fatty acids		Polyene fatty acids	
Fatty acid	Weight %	Fatty acid	Weight %	Fatty acid	Weight %
10:1	0.48	14:2	0.04	18:3	1.03
12:1	0.05	16:2	0.02	18:4	0.10
13:1	0.003	18:2	2.22	20:3	0.05
14:1	0.75	20:2	0.12	20:4	0.07
15:1	0.02	22:2	0.14	20:5	0.02
16:1	1.84	24:2	0.02	22:3	0.03
17:1	0.2	26:2	0.0004	22:4	0.04
18:1	30.3			22:5	0.02
19:1	0.14				
20:1	0.52				
21:1	0.01				
22:1	0.02				
23:1	0.05				
24:1	0.0008				
25:1	0.0008				
26:1	0.0008				

Values obtained from Iverson *et al.*, 1965.

the satisfactory columns now available, there are problems in GC triglyceride analysis. Because of the high molecular weight, injection usually has to be on-column at a low temperature, or using a programmed (PVT) injector (Banfi and Bergna, 1999), while it is particularly important that dead-volume between column and injector is minimal. High temperature columns are produced that are more polar than those normally used in triglyceride analysis and that claim to give more separating power, and thus separate more peaks. These can cause losses of more unsaturated triglycerides, and, more importantly in the case of milk fat, produce a more complicated picture, which does not help much in milk fat authentication. GC of milk fat is probably best carried out using a non-polar column, and thus merely separating the fat by carbon number, and not attempting further separation.

Triglyceride composition, as determined by measuring the carbon numbers of the triglyceride fraction, is affected by many of the same factors as is fatty acid composition. Canola oil feeding increases C50, C52 and C54 triglycerides, and reduces C32, C34 and C36 (DePeters *et al.*, 2001). This might be expected, as virtually all the triglycerides of rapeseed are in that first group. Feeding of special diets designed to increase spreadability, by including high levels of unsaturated C18 fatty acids, reduced saturated triglycerides, especially for carbon numbers C42–C50. Considerable differences are also found between summer and winter milk fat compositions in Germany (Hinrichs *et al.*, 1992), and between fat from cattle in high and lowland regions (Collomb *et al.*, 1999). Table 5.5 lists the ranges of triglyceride compositions reported for genuine milk fats, including those fed unusual, but commercially likely, diets. Early attempts to detect foreign fats from triglyceride analyses using regression procedures could theoretically detect levels of 4–7% of some oils, but could not handle mixtures of oils (Lercker *et al.*, 1992; Luff, 1987; Luff *et al.*, 1987; Luff, 1988a,b), while examination of the trisaturated triglyceride fraction for triglyceride composition was even less indicative of adulteration (Parodi and Dunstan, 1969). Precht later developed from the analysis of 755 milk fat samples a statistically derived formulae for triglyceride composition of milk fat, which he claimed enabled soyabean, sunflower, olive, coconut, palm and palm kernel oils and lard and beef tallow to be detected with 95% confidence at levels of detection of (respectively) 1.3, 1.2, 1.7, 2.5, 2.7, 2.9, 1.9 and 4.1% (Precht and Heine, 1986; Precht, 1991, 1992a,b). The method was tested and found satisfactory at about the detection level of the foreign fats and with this range of adulterants. It was pointed out that, where more than one fat was present in the product, these detection levels were not appropriate. Where mixtures of fats are present then detection levels are around 4–7%. When the method was used to attempt to detect non-dairy fat in Italian cheeses, it was found to be successful for those cheeses that had not undergone extensive lipolysis, but indicated falsely the presence of foreign fats when lipolysis was significant. This was attributed to loss of low molecular weight triglycerides, but the explanation was only tentative. Other workers have

Table 5.5 Range of values reported for triglyceride composition of genuine milk fats by carbon number gas chromatography

Triglyceride carbon number	Minimum	Maximum
24	0.00	0.11
26	0.14	0.76
28	0.40	0.82
30	0.56	1.63
32	1.44	3.76
34	3.86	7.90
36	8.21	13.81
38	10.00	14.36
40	8.35	12.04
42	4.95	8.65
44	4.05	8.89
46	4.88	9.71
48	6.80	10.65
50	10.13	15.08
52	5.13	15.49
54	1.54	11.09
56	0.11	0.41

Carbon numbers of triglycerides obtained by adding the carbon numbers of the component fatty acids, ignoring those of glycerol. Values obtained from a compilation of a number of results in the literature.

tested the equations and found them satisfactory for butters from other sources (Collomb *et al.*, 1998a,b).

The official European Community (EC) method for determination of the purity of milk fat is based on triglyceride composition as determined by GC on a packed column (EC, 1999). This is based on a formula for the composition of genuine milk fat of the type: $100 = 14.197\ C40 - 36.396\ C42 + 32.364\ C44 - e$. The limit of detection varies with the adulterant fat, but is usually <5% foreign fat. A standard reference material (CRM 519) is available (Precht *et al.*, 1998) which has been fully characterized and the triglyceride composition determined collaboratively. Other methods of analysis of the triglyceride data have been evaluated (Collomb *et al.*, 1998a,b; Lipp, 1996a,b; Ulberth, 1995).

It has been suggested (Contarini *et al.*, 1999; Povolo *et al.*, 1999) that, in order to lower the detection limit for beef tallow in milk fat, statistical evaluation of the data for triglycerides with data for diglycerides and cholesta-3,5-diene analysis should be carried out using the UNEQ technique. Unrefined tallow or lard would probably give an unpleasant odour to the milk fat, and so refining is probably necessary if these are incorporated. This is likely to cause the formation of cholesta-3,5-diene (Bianchi *et al.*, 1996; Kuzdzal-Savoie *et al.*, 1975; Roderbourg and Kuzdzal-Savoie, 1979). Diglycerides of tallow (containing mainly C14:0, C16:0, C18:0, C18:1, C18:2) are somewhat different in composition to milk fat triglycerides, where a more complex array of fatty acids is present. Thus incorporation of these results into the analysis enabled detection of beef tallow down to 2%.

Milk fat triglycerides can also be determined by HPLC analysis. This involves separation on a C18-column, or possibly two columns in series, with elution using a variety of polar solvent mixtures, possibly using a solvent gradient (Frede and Thiele, 1987; Maniongui *et al.*, 1991; Mottram, 1999; Najera *et al.*, 1999; Robinson and Macgibbon, 1998; Spanos *et al.*, 1995). The most satisfactory method of detection is by the light scattering detector, though gas chromatography/mass spectrometry has also been used. It is likely, when a standard method has been evaluated, that the greater separation of components might enable this method of evaluation to provide an even lower detection limit for adulterants of milk fat. Before this occurs it would be necessary to provide a satisfactory, reproducible method, and to investigate the ranges of composition obtainable from butters of different origin. Up till now only individual fats have been analysed, whereas Precht (1991, 1992a,b) analysed 755 milk samples for GC carbon number triglyceride composition before developing his equations. The data presented from a similar number of HPLC analyses would undoubtedly need a great deal of data processing, but nowadays this is not a great problem. It should be noted that there would be no need to identify any of the peaks observed, indeed many will contain more than one component. All that would be necessary would be to carry out the analysis using an agreed reproducible separation.

5.3.1.4 Other methods of milk fat analysis

Several other methods have been suggested for the detection of foreign fats in milk fat (Collomb and Spahni, 1991). Near-infra red analysis (NIRA) has been proposed (LaPorte and Paquin, 1999), but, as with all use of NIRA, accurate results depend on a good and comprehensive data set of accurately pre-analysed samples that cover the whole range to be encountered. While it might be useful as a quick check it would not be likely to be useful generally. The use of determination of squalene and octadecanol has also been suggested (Maritano de Correche and Oxley, 1985), but has not been further investigated. ^{13}C nuclear magnetic resonance (NMR) spectroscopy (Andreotti *et al.*, 2002) has been suggested as a tool to the authentication of milk fat because of differences in the positional distribution of fatty acids on the glycerol, this making the results dependent on the same factors as utilizing lipase analysis of the acid at the 2-position, but much quicker. Finally, differential scanning calorimetry has been proposed (Amelotti *et al.*, 1983; Coni *et al.*, 1994; Sato, 1997; Tunick *et al.*, 1998), but, even in known spiked samples, was shown to detect no lower levels of adulteration than 5–10%.

5.3.1.5 *Refined and modified milk fat*

Although one might consider the most desirable property of milk fat is the flavour, for some purposes refining, hydrogenation and fractionation are carried out on the fat. As has previously been stated, refining with, for example, fullers earth or another similar bleaching earth, gives rise to the formation of cholesta-3,5-diene, which may be determined (Huyghebaert and de Moor, 1974; Kuzdzal-Savoie *et al.*, 1975; Roderbourg and Kuzdzal-Savoie, 1979) if required. Fractionation will give rise to a different triglyceride and fatty acid composition which should be shown up by the standard methods of triglyceride analysis, the results depending on whether it is a soft or a hard fraction that is present. Cholesterol levels will also be altered on fractionation, it being concentrated into the more liquid fractions (Arul *et al.*, 1988). Hydrogenation, in addition to hydrogenating the fatty acids, also hydrogenates hydrocarbons. Thus hydrogenated milk fats contain squalane and other products not found in the unhydrogenated fat and derived from squalene (Kuzdzal-Savoie *et al.*, 1975), while, if the fat is refined with bleaching earth and then hydrogenated, cholestene is found to be formed (Roderbourg and Kuzdzal-Savoie, 1979). It has also been reported (Eisner *et al.*, 1966) that hydrogenation of milk fat forms (Roderbourg and Kuzdzal-Savoie, 1979) an extra component tentatively identified as a C30-branched alcohol. These facts should make modified milk fats easily detectable in milk fat.

5.3.2 *Milk fat from other animal sources*

Recently it has been claimed that milks from cow, buffalo, goat and sheep can be differentiated quickly by ^{13}C NMR spectroscopy (Andreotti *et al.*, 2002), though a large data set has not yet apparently been evaluated. This separation is apparently possible because of the triglyceride composition of the milks. If it is possible that any DNA is present in the product, this being very likely with butter or many milk products, then DNA analysis can detect cow milk in ewe, goat, and buffalo with a sensitivity of about 5% (Klotz and Einspanier, 2001). Similarly, where protein is present, capillary electrophoresis is claimed to detect down to 1% of cow milk in goat milk, and down to 7% in goat cheese (Ramos *et al.*, 1977). A survey of what is possible using similar techniques in detecting cow milk in goat and sheep cheeses is given by Uotila (1996).

With regard to the fat components of milks, although comparatively little work has been carried out on variations in composition of milk other than that from cattle, it is likely that similar variations occur. Sheep's milk varies considerably with season (Gattuso and Fazio, 1980; Matter, 1988; Perea *et al.*, 2000), with regard to degree of unsaturation, amount of long chain fatty acids and triglyceride composition. This has been associated with differences of diet. Similar variations occur over the year with goat milk (Fontecha *et al.*, 1998), and undoubtably also occurs with the milk fat from other animals. The ranges

Table 5.6 Ranges of composition of major fatty acids reported in cheeses manufactured from milk of different species

	Cows		Buffalo		Sheep		Goats	
Fatty acid	Min.	Max.	Min.	Max.	Min.	Max.	Min.	Max.
C8:0	0.72	1.40	0.57	1.05	1.31	2.72	2.07	2.78
C10:0	2.20	3.37	1.24	2.38	3.76	8.20	7.53	9.89
C12:0	2.63	4.22	1.39	3.08	2.53	5.34	3.48	4.77
C14:0	9.31	13.3	8.41	12.4	6.57	12.5	8.09	11.6
C16:0	22.6	36.1	26.4	34.5	13.8	28.9	22.2	31.4
C18:0	6.80	13.2	9.00	18.0	6.83	13.7	7.56	12.6
C18:1	19.9	28.8	18.3	31.8	18.4	28.1	18.5	25.1

Lower fatty acids omitted. Adapted from Prager (1989).

of fatty acids found in fat from cheeses manufactured from cows, buffalo, sheep and goat milks are listed by Prager (1989), and given in Table 5.6. He proceeded to calculate ratios of fatty acids which could be used to check for adulteration with milk from another source than would be expected. Lower fatty acids were omitted from the calculation due to variability. Iverson and Sheppard (1989) came to similar conclusions, but concentrated on the ratio C12/C10 only, this figure showing the most difference between species. Their ratios differed slightly from Prager's, and they did not consider buffalo milk.

The pattern of fatty acids in milk from animals other than cattle does differ from bovine milk (Hilditch and Williams, 1964; Parodi, 1973a,b). As has already been stated, the even-numbered fatty acids of bovine milk tend to increase approximately with increasing carbon number. This is not necessarily true of other species. Goat and sheep milks tend to have lower amounts of C12:0 than C10:0, though this might not be true if the animal had been fed on a diet containing significant levels of coconut or palm kernel oil. This scenario is not, however, very likely to occur. With the exception of buffalo milk, non-bovine milks tend to have lower levels of butyric acid than bovine milk. These differences are carried over into the triglyceride compositions. GC and HPLC analysis show that ewe's milk is richer than bovine or goat milk in short and medium length triglycerides (Ruiz-Sala *et al.*, 1996), while both goat and ewe's milk are lower in higher molecular weight triglycerides (Fontecha *et al.*, 2000). This results in the distribution of triglycerides in goat milk fat being unimodal, peaking at C40, unlike bovine milk which is bimodal.

Use of triglyceride analysis, together with multiple regression equations has been claimed to be successful in determining adulteration of goat milk with bovine milk fat or other adulterant fats (Fontecha *et al.*, 1998; Luff, 1987; Luff *et al.*, 1987). Buffalo milk fat adulterated by bovine milk fat has been successfully detected by fractionation of the fat, followed by fatty acid analysis and regression analysis (Farag *et al.*, 1980; Farag and Hewedi, 1984, 1986), and adulteration with lard can be detected by fatty acid analysis, and calculation of acid ratios.

5.4 Carcass fats

5.4.1 Beef tallow

After the possibility of vegetable fats has been eliminated by testing for plant sterols, as with milk fats, then the fatty acid composition of the fat would be the next approach in testing beef tallow. The accepted fatty acid composition range for beef tallow and premier jus, according to *Codex Alimentarius* standards, is given in Table 5.2. Premier jus is mainly distinguished from edible beef tallows by the free fatty acid (FFA) level of <1%, and is used in premier products such as suet. Edible tallow (FFA = 1–1.25%) can also, if declared, contain mutton fat (Rossell, 2001), but not if it is not declared. The major fatty acids are C16:0, C18:0 and C18:1. If the level of contamination is fairly high then simple fatty acid analysis will suffice to detect most adulteration, including pork (Javidipour *et al.*, 1999; Matter, 1991; Wurziger and Hensel, 1969; Verbeke and de Brabander, 1980) and possibly horse fat (Verbeke and de Brabander, 1980). The presence of vegetable fats may also be detected.

The specific presence of eicosa-11,14-dienoic acid, a component of pork fat but not of beef or mutton, has been suggested as a test for the presence of lard (Saeed *et al.*, 1986), but this would only be of use if vegetable fats were not present as these sometimes also contain this acid in small amounts. Similarly ozonolysis of the triglycerides followed by HPLC, in order to detect and separate higher levels of unsymmetrical triglycerides with the saturated acids at the 2-position, is stated to detect down to 1% pork fat in beef tallow (Saeed *et al.*, 1989), but natural variations in composition would undoubtably increase this detection limit, while the presence of vegetable fats is likely to mask the effect of the pork fat. A similar separation in which similar results for animal fat mixtures could probably be obtained uses halogen addition and HPLC, and has been used for investigations of vegetable oils (Podlaha and Toregard, 1984). Also relying on the unsymmetrical triglycerides of pork fat is the determination of enrichment factors for beef and pork fats of palmitic acid at the 2-position over the level in the total triglycerides (<0.8 for beef, >0.8 for pork), together with unsaturation ratios, these all being obtained after lipase hydrolysis (Dabash *et al.*, 1979). Vegetable fats might interfere. It is also possible that trace fatty acids in tallow, separated by urea fractionation, might enable identification of adulterants, both animal and vegetable (Iverson and Weik, 1967; Peters and Wieske, 1966). After a long period of relative disuse this process of concentration has recently been further investigated, mainly for production purposes (Hayes *et al.*, 1998). It is possible that the process might also find further uses in analysis.

5.4.2 Pork fat

As with beef tallow, after elimination of the possibility of the presence of vegetable fats, the presence of levels of relatively high levels of adulterant fats can usually be detected by fatty acid analysis. The *Codex Alimentarius* recommended acceptable fatty acid compositions for pork fat are given in Table 5.2. Unusual fatty acid patterns can indicate the presence of foreign fats (Hubbard and Pocklington, 1968), while enrichment ratios at the 2-position of the triglyceride of palmitic acid and of unsaturation can also be used for this purpose (Dabash *et al.*, 1979; Podlaha and Toregard, 1984; Verbeke and de Brabander, 1980), as can derivatisation of triglycerides followed by HPLC (Podlaha and Toregard, 1984; Saeed *et al.*, 1989). It has been suggested that the presence of some branched chain fatty acids in pork fat are an indication of adulteration with a ruminant fat such as beef or mutton tallows (Hubbard and Pocklington, 1968; Juarez and Martinez-Castro, 1981; Lotito and Cucurachi, 1966). While this a is likely indication, it is known (Bastijns, 1968, 1970) that feeding of beef, and presumably mutton, fat to pigs can cause levels of these acids to increase in the lard from the pig. Therefore any conclusion based on this evidence should preferably be confirmed, though it is appreciated that, since BSE in cattle, feeding of beef to pigs may be unlikely. Possibly urea fractionation of all the minor saturated branched fatty acids (Iverson and Weik, 1967; Peters and Wieske, 1966) could be investigated as an additional source of information. Little work has been reported on HPLC of triglycerides but HPLC traces of lard and beef tallow triglycerides do show great differences (Perrin and Prevot,

Table 5.7 Comparison of typical fatty acid compositions (weight %) of subcutaneous adipose tissue from several animals

Fatty acid	Beef	Sheep	Pig	Horse	Chicken	Duck	Goose
C10:0	0–0.1	0.1–0.2	0.1	—	—	—	—
C12:0	0.1	0.1–0.5	0.1	—	—	—	—
C14:0	2.7–4.8	2.8–4.9	1.4–1.7	4.0	1.3	0.7	0.5
C14:1	0.8–2.5	0.7–0.8	0–0.1	0.4	0.2	—	—
C16:0	20.9–28.9	19.5–21.3	23.1–28.3	27.9	23.2	25.9	22.0
C16:1	2.3–9.1	1.4–2.3	1.8–3.3	8.0	6.5	4.2	3.0
C17:0	1.0	1.0	0.5	0.3	0.3	—	—
C18:0	7.0–26.5	17.6–28.9	11.7–24	4.2	6.4	8.2	6.5
C18:1	30.4–48.0	33.2–40.4	29.7–45.3	34.0	41.6	46.3	56.9
C18:2	0.6–1.8	1.2–3.4	8.1–12.6	6.7	18.9	12.6	10.4
C18:3	0.3–0.7	1.4–1.9	0.7–1.2	10.1	1.3	1.0	0.5
C:20:0	tr–0.9	tr–0.3	0.2–0.3	0.1	—	—	—
C20:1	0.3–1.7	0.2–0.3	0.8–1.3	—	—	—	—
C20:2	0–0.1	0	0.3–0.5	0.3	—	—	—
C20:4	—	—	—	4.5	—	—	—
C22:0	0–0.1	0	0–0.4	—	—	—	—
C22:1	0–tr	0	tr–0.1	—	—	—	—
C24:0	0–tr	0	0–0.5	—	—	—	—
C24:1	0	0	0–0.6	—	—	—	—
trans-	1.3–6.6	11.0–14.6	1.1–1.4	—	—	—	—

Adapted from Rossell, 2001.
tr, <0.05%.

1986). It is likely that this could also be useful in determining adulteration, but more data have to be collected before this can be confirmed.

5.4.3 Authentication of fat from other sources

Relatively little work has been carried out on adulteration of animal carcass fats other than beef or pork fat. Table 5.7 (Rossell, 2001) compares typical compositions from some species. Certain differences, such as the presence of high levels of C20:4 in horse fat, and the high levels of C18:2 in the fats from chicken, goose and duck, are the most obvious. From fatty acid compositions, adulteration of goose fat with 10% lard (Brixins and Treiber, 1974; Matter, 1991, 1992) can be determined from stearic/oleic acid ratios, and similar ratios can be used to detect mixtures of chicken and turkey fat (Matter, 1991). Lard in buffalo tallow can be detected by a combination of C18:0/C18:2 ratios, and the palmitic acid enrichment factor of 2-monoglycerides over the whole triglyceride (Youssef and Rashwan, 1987), as can lard in goat and mutton tallows (Rashwan and Youssef, 1989). The fatty acid composition can also differentiate between feral and domestic pigs and between feral hares and domestic rabbits by respectively the ratios C17:0/C16:0 and C18:2/C18:3 (Matter, 1992).

As a quick method of detection it has been suggested that differential scanning calorimetry (DSC) can detect lard adulteration in goat and mutton tallows, though the level of detection is not likely to be very low.

5.5 Conclusions

Sterol analysis, fatty acid analysis of the whole fat and of the acids at the 2-position, and triglycerides by GC, can be very useful in determining adulteration of and by animal fats. In the future HPLC of triglycerides is likely to provide an even better method for some purposes, but much more data need to be collected before this can be evaluated.

References

Akerlind, M., Holtenius, K., Bertilsson, J. and Emanuelson, M. (1999) Milk composition and feed intake in dairy cows selected for high or low milk fat percentage. *Livestock Prod. Sci.*, **59**, 1–11.

Amelotti, G. Brianza, M. and Lodigiani, P. (1983) Applications and limits of differential scanning calorimetry to the detection of beef suet in butter. *Riv. Ital. Sost. Grasse*, **60**, 557–564.

Andreotti, G., Lamanna, R., Trivellone, E. and Motta, A. (2002) ^{13}C NMR spectra of TAG: An easy way to distinguish milks from different animal species. *J. Am. Oil Chem. Soc.*, **79**, In Press.

AOAC (Association of Official Analytical Chemists) (1995a) AOAC Official Method 955.34 Fats (Vegetable) in Butterfat, in *Official Methods of Analysis of the Association of Official Analytical Chemists*, 16th edn, Arlington, Virginia, USA.

AOAC (Association of Official Analytical Chemists) (1995b) AOAC Official Method 970.51 Fats (Animal) in Vegetable Fats and Oils (Determination of Cholesterol), in *Official Methods of Analysis of the Association of Official Analytical Chemists*, 16th edn, Arlington, Virginia, USA.

AOAC (Association of Official Analytical Chemists) (1995c) AOAC Official Method 970.50 Fats (Vegetable) in Butterfat, in *Official Methods of Analysis of the Association of Official Analytical Chemists*, 16th edn, Arlington, Virginia, USA.

AOAC (Association of Official Analytical Chemists) (1995d) AOAC Official Method 967.18 Beta-Sitosterol in Butteroil, in *Official Methods of Analysis of the Association of Official Analytical Chemists*, 16th edn, Arlington, Virginia, USA.

Arul, J., Bordreau, A., Makhlouf, J., Tardif, R. and Grenier, B. (1988) Distribution of cholesterol in milk fat fractions. *J. Dairy Res.*, **55**, 361–371.

Ballesteros, E., Gallego, M. and Valcarcel, M. (1995) Simultaneous determination of sterols in edible oils by use of a continuous separation module coupled to a gas chromatograph. *Anal. Chim. Acta.*, **308**, 253–260.

Banfi, S. and Bergna, M. (1999) Programmable temperature vaporizer (PVT) applied to the triglyceride analysis of milk fat. *J. High Resol. Chrom. Commun.*, **22**, 93–96.

Bastjns, L.J. (1968) The influence of diet on the composition of the fatty acids of lard. *Chem. Indus.*, 721–722.

Bastjns, L.J. (1970) Composition of lard fatty acids from pigs given beef fats in their diet. *J. Sci. Food Agric.*, **21**, 576–578.

Battelli, G. and Pellegrino, L. (1995) Detection of non-dairy fat in cheese by gas chromatography of triglycerides. *Ital. J. Food Sci.*, **6**, 407–419.

Beaulieu, A.D. and Palmquist, D.L. (1995) Differential effects of high fat diets on fatty acid composition in milk of Jersey and Holstein cows. *J. Dairy Sci.*, **78**, 1336–1344.

Bianchi, E., Fabietti, F., Delise, M., Cantoni, A. and Zambini, E.M. (1996) Analytical profiles to distinguish refined and unrefined lards with particular regard to the 3,5-cholestadiene content, *Indus. Aliment.*, **35**, 1082–1086.

Bonvehi, J.S., Torrento, M.S. and Coll, F.V. (2001) A laboratory study of the bleaching process in stigmasta-3,5-diene concentration in olive oils., *J. Am. Oil Chem. Soc.*, **78**, 305–310.

Brixins, L. and Treiber, H. (1974) Comparative gas chromatographic studies on the fatty acid composition of lard and goose dipping. *Fette Seif. Anstrichm.*, **76**, 83–86.

Camin, R., Scheufler, E., Blaschka, A., Jahn, S., Sparborth, D., Franke, H. and Doebel, K. (2000) Process for manufacture of cholesterol reduced fats and corresponding sausages made from pork and pork fat. German Patent application DE19917929A1.

Caponio, F., Allogio, V. and Pallavicini, C. (1998) Modification of goat milk fat triglycerides by immobilised lipase. *Fett Lipid*, **100**, 74–78.

Cert, A., Lanzon, A., Carelli, A.A., Albi, T. and Amelotti, G. (1994) Formation of stignasta-3,5-diene in vegetable oils. *Food Chem.*, **49**, 283–287.

Chilliard, Y., Ferlay, A., Mansbridge, R.M. and Doreau, M. (2000) Ruminant milk fat plasticity: nutritional control of saturated, polyunsaturated, trans and conjugated fatty acids. *Ann. Zootech.*, **49**, 181–205.

Chung, W.C. and Lucy, S.H. (1996) Determination of free and esterified cholesterol contents in lard, *Food Sci. Taiwan*, **23**, 367–376.

Codex Alimentarius Commission (1993a) Codex Standard for premier jus 30-1981 (Rev.1-1989), *Codex Alimentarius*, vol. 8, *Fats, oils and related products*, FAO/Rome, p. 87.

Codex Alimentarius Commission (1993b) Codex Standard for edible tallow, 31-1981 (Rev.1-1989), *Codex Alimentarius*, vol. 8, *Fats, oils and related products*, FAO/Rome, p. 91.

Codex Alimentarius Commission (1993c) Codex Standard for lard 28-1981 (Rev.1-1989), *Codex Alimentarius*, vol. 8, *Fats, oils and related products*, FAO/Rome, p. 79.

Codex Alimentarius Commission (1997) Proposed Draft Codex Standard for Named Vegetable Oils, FAO/Rome, p. 79.

Collomb, M. and Spahni, M. (1991) Adulteration of milk products. Review of the analytical criteria for the detection of vegetable and animal fats in milk fat. *Mitt. Gebiete Lebensm. Hugg.*, **82**, 615–662.

Collomb, M., Buetikofer, U., Spahni, M. and Buehler, T. (1998a) Analysis of triglycerides. Comparison of methods of MLR, PLS and PCA regression for the determination of foreign fats in milk fat using capillary GC. *Mitt. Gebiete Lebensm. Hyg.*, **89**, 617–624.

Collomb, M., Spahni, M. and Buehler, T. (1998b) Analysis of triglycerides. 2. Validation of a multilinear regression method for the determination of foreign fats in milk fat. *Mitt. Gebiete Lebensm. Hyg.*, **89**, 75–83.

Collomb, M., Buetikofer, U., Spahni, M., Jeangros, B. and Bosset, J.O. (1999) Fatty acid and glyceride composition of cow's milk fat in high- and lowland regions. *Sci. Aliment.* **19**, 97–110.

Coni, E., Di-Pasquale, M., Coppolelli, P. and Bocca, A. (1994) Detection of animal fats in butter by differential scanning calorimetry. *J. Am. Oil Chem. Soc.*, **71**, 807–810.

Contarini, G., Povolo, M. and Bonfitto, E. (1999) The present and future of methods for controlling the authenticity of butter. *Latte*, **24**, 60–69.

Dabash, A.S., Gebriel, A.Y., Hussein, M.F. and El-Dashlouty, A.A, (1979) Detection of lard in tinned meat. *Res. Bull. Fac. Agric., Ain Shams Univ., No 1043*, Pub. Ain Shams Univ., Cairo.

DePeters, E.J., German, J.B., Taylor, S.J., Essex, S.T. and Perez-Monti, H. (2001) Fatty acid and triglyceride composition of milk fat from lactating Holstein cows in response to supplemental canola oil. *J. Dairy Sci.*, **84**, 929–936.

Drackley, J.K., Beaulieu, A.D. and Elliott, J.P. (2001) Responses of milk fat composition to dietary fat or nonstructural carbohydrates in Holstein and Jersey cows. *J. Dairy Sci.*, **84**, 1231–1237.

EC (European Community) (1991) EC Council Regulation 2568/91 Characteristics of olive oil and olive residue oil and on the relevant methods of analysis.

EC (European Community) (1999) EC Council Regulation No. 2771/1999, Official Journal L333. 24/12/1999, pp. 11–43.

Eisner, J., Iverson, J.L. and Firestone, D. (1966) Gas chromatography of unsaponifiable matter. 4. Aliphatic alcohols, tocopherols and triterpenoid alcohols in butter and vegetable oils. *J. Assoc. Off. Anal. Chem.*, **49**, 580–590.

Farag, R.S. and Hewedi, M. (1984) Detection of cows milk admixture to buffalo's milk. *J. Am. Oil Chem. Soc.*, **61**, 913–916.

Farag, R.S. and Hewedi, M. (1986) Detection of cows milk admixture to buffalo's milk. *Fette Seif. Anstrichm.*, **88**, 106–108.

Farag, R.S., Abl El Samad, A. and El-Rafey, H.H.A. (1980) Detection of lard and shortening adulteration in pure buffalo and cow ghee. *Res. Bull. Fac. Agric., Ain Shams Univ., No 1283*. Pub. Ain Shams Univ., Cairo.

Farag, R.S., Ahmed, F.A., Shihata, A.A., Abo-raya, S.H. and Abdalla, A.F. (1982) Use of unsaponifiable matter for detection of ghee adulteration with other fats. *J. Am. Oil Chem. Soc.*, **59**, 557–560.

Farag, R.S., Abo-Raya, S.H., Ahmed, F.A., Hewedi, F.M. and Khalifa, H.H. (1983) Fractional crystallization and gas-chromatographic analysis of fatty acids as a means of detecting butter-fat adulteration. *J. Am Oil Chem. Soc.*, **60**, 1665–1669.

Fearon, A.M., Charlton, C.T. and Kilpatrick, D.J. (1994) A further investigation of the influence of dietary protected lipid supplements on the characteristics of cow's milk fat. *J. Sci. Food Agric.*, **66**, 247–256.

Firestone, D. (2001) Assuring the integrity of olive oil products. *J. Assoc. Off. Anal. Chem.*, **84**, 176–180.

Fontecha, J., Diaz, V., Fraga, M.J. and Juarez, M. (1998) Triglyceride analysis by gas chromatography in assessment of authenticity of goat milk fat. *J. Am. Oil Chem. Soc.*, **75**, 1893–1896.

Fontecha, J., Rios, J.J., Lozada, L., Fraga, M.J. and Juarez, M. (2000) Composition of goat's milk fat triglycerides analysed by silver ion absorption-TLC and GC-MS. *Int. Dairy J.*, **10**, 119–128.

Fox, J.R., Duthie, A.H. and Wulff, S. (1989) Effect of seasonal variation of fatty acids in milk fat on the sensitivity of a test for vegetable oil adulteration. *J. Dairy Sci.*, **72**, 1981–1985.

Frede, E. and Theile, H. (1987) Analysis of milk fat by HPLC. *J. Am. Oil Chem. Soc.*, **64**, 521–528.

Gattuso, A.M. and Fazio, G. (1980) Characteristics and composition of milk fat from sheep of Sicilian breeds. *Riv. Ital. Sost. Grasse*, **57**, 530–535.

German, J.B., Simoneau, C., DePeters, E., Tong, P.S., Mattas, J. and Sullivan, J. (1995) Effects of feeding supplemental fat to dairy cattle on milk fat composition and properties. 2. Triglyceride distribution and milk fat functionality. *J. Dairy Sci.*, **78**, (Suppl. 1), 134.

Gomer, P. and Kotulak, R. (1991) Gene splicers putting new food on the table. *Food Technol.*, **45**, 46,48,53,103.

Gow, C.Y. and Li, J.T. (1995) Cholesterol removal from a lard-water mixture with beta-cyclodextrin, *J. Food Sci.*, **60**, 561–564.

Grob Jr, K. (1984) Effect of 'dirt' injected on-column in capillary gas chromatography; analysis of the sterol fraction of oils as an example. *J. Chrom.*, **287**, 1–14.

Grob, K., Giuffre, A.M., Leuzzi, B. and Mincione, B. (1994) Methoden zur Erkennung verfaischter Speiseole. *Mitt. Gebiete Lebensm. Hyg.*, **85**, 340–350.

Gu, Y.F., Chen, Y. and Hammond, E.G., (1994) Use of cyclic anhydrides to remove cholesterol and other hydroxy compounds from fats and oils. *J. Am. Oil Chem. Soc.*, **71**, 1205–1209.

Gunstone, F.D., Harwood, J.I. and Padley, F.B. (1994) *The Lipids Handbook*, 2nd edn, Chapman & Hall.

Guyot, A. (1969) Identification of fat mixtures by analysis of fatty acids and sterols. *Bull. Rech. Agron. Gembloux*, **4**, 484–508.

Hadorn, H. and Zurcher, K. (1972) Butterfat determination in chocolate by gas chromatography. *Rev. Int. Chocolate*, **27**, 82–92.

Hagemeister, H., Precht, D., Franzen, M. and Barth, C.A. (1991) Alpha-linoleic acid transfer into milk fat and its elongation by cows. *Fett Wissen. Technol.*, **93**, 387–391.

Hallabo, S.A. and El-Nikeety, M.M. (1987) Comparative study between imported butter oil and Egyptian cow ghee. *Egypt. J. Food Sci.*, **15**, 195–202.

Hayes, D.G., Bengtsson, Y.C.,Van Alstine, J..M. and Setterwell, F. (1998) Urea complexation for the rapid, ecologically responsible fractionation of fatty acids from seed oil. *J. Am. Oil Chem. Soc.*, **75**, 1403–1409.

Hermansen, J.E. (1995) Prediction of milk fatty acid profile in dairy cows fed dietary fat differing in fatty acid composition. *J. Dairy Sci.*, **78**, 872–879.

Hettinga, D. (1996) Butter, in *Bailey's Industrial Oil and Fat Products*, 5th edn (ed. Y.H. Hui), John Wiley and Sons, New York, p. 1.

Hilditch, T.P. & Williams, P.N. (1964) *The Chemical Composition of Natural Fats.*, 4th edn, Chapman & Hall, London.

Hinrichs, J., Heinemann, U. and Kessler, H.G. (1992) Differences in the composition of triglycerides in summer and winter milk fat. *Milchwissenschaft*, **47**, 495–498.

Hodd, T. (1996) Kosher and non-kosher food emulsifiers. *Int. Food Ingredients*, **5**, 31–32, 35–37.

Homberg, E. (1991) Analysis of sterols as a means for detection of blended and adulterated fats. *Fett Wissen. Technol.*, **93**, 516–517.

Homberg, E. and Bielefeld, B. (1987) Comparative determinations of sterols in fats and oils. *Fett Wissen. Technol.*, **89**, 255–258.

Hubbard, A.W. and Pocklington, W.D. (1968) Distribution of fatty acids in lipids as an aid to the identification of animal tissues. 1. Bovine, ovine and some avian species. *J. Sci. Food Agric.*, **19**, 571–577.

Huyghebaert, A. and de Moor, H. (1974) The influence of some refining processes on the sterols of butter oil. *Proceedings of the 14th International Dairy Congress.*, **1E**, 215–216.

Iverson, J.L. and Sheppard, A.J. (1986) Determination of fatty acids in butter fat using temperature-programmed gas chromatography of the butyl esters. *Food Chem.*, **21**, 223–234.

Iverson, J.L. and Sheppard, A.J. (1989) Detection of adulteration in cow, goat, and sheep cheeses utilizing gas-liquid-chromatographic fatty acid data. *J. Dairy Sci.*, **72**, 1707–1712.

Iverson, J.L. and Weik, R.W. (1967) Correlation of fatty acid structure with preferential order of urea complex formation. *J. Assoc. Off. Anal. Chem.*, **50**, 1111–1118.

Iverson, J.L., Eisner, J. and Firestone, J. (1965) Fatty acids of milk fat. *J. Am. Oil Chem. Soc.*, **42**, 1063–1068.

Javidipour, I., Tekin, A. and Ergin, G. (1999) Determination of lard in adulterated beef tallow by gas chromatography. *Gida*, **24**, 171–175.

Jenkins, T.C. (1998) Fatty acid composition of milk from Holstein cows fed oleamide or canola oil. *J. Dairy Sci.*, **81**, 794–800.

Jones, D.F., Weiss, W.P. and Palmquist, D.L. (2000) Short communication: influence of dietary tallow and fish oil on milk fat composition. *J. Dairy Sci.*, **83**, 2024–2026.

Juarez, M. and Martinez-Castro, I. (1981) Detection of suet in lard by GC of fatty acids. *Grasas Aceitas*, **32**, 93–96.

Kanematsu, H., Maruyama, T., Niiya, I., Amamura, A. and Matsumoto, T. (1973) Changes in the steroid content of edible fats and oils during commercial processing. *J. Japan Oil Chem. Soc.*, **22**, 814–817.

Karleskind, A. (1969) Study of unsaponifiable alcohols. 3. Application to the study of complex mixtures. *Rev. Franc. Corps. Gras.*, **16**, 15–24.

Kiosseoglou, B. and Boskou, D. (1987) On the level of esterified sterols in cottonseed, tomato seed, wheat germ and safflower oils. *Oleagineux*, **42**, 169–170.

Klotz, A. and Einspanier, R. (2001) Development of a DNA-based screening method to detect cow milk in ewe, goat and buffalo milk and dairy products using PCR-LCR-EIA technique. *Milchwissenschaft*, **56**, 67–70.

Kuzdzal-Savoie, S., Langlois, D., Trotier, D. and Dzik, B. (1975) Comparative study of some hydrocarbons in butter and modified butter. *Ann. Falsifications et del'Expertise Chim.*, **68**, 577–617.

LaCroix, D.E. (1970) Collaborative study of the extraction of plant sterols from adulterated butter oil using a digitonin-impregnated Celite column. *J. Assoc. Off. Anal. Chem.*, **53**, 535–538.

Lanzani, A., Bondioli, P., Mariani, C., Folegatti, L., Venturini, S., Fedeli, E. and Barreteau, P. (1994) A new short-path distillation system applied to the reduction of cholesterol in butter and lard. *J. Am. Oil Chem. Soc.*, **71**, 609–614.

Lanzon, A., Albi, T., Cert, A. and Gracian, J. (1994) The hydrocarbon fraction of virgin olive oil and changes resulting from refining. *J. Am. Oil Chem. Soc.*, **71**, 285–291.

LaPorte, M.F. and Paquin, P. (1998) Near-infrared technology and dairy food products analysis: a review. *Semin. Food Anal.*, **3**, 173–190.

Lercker, G., Frega, N., Bocci, F. and Bertacco, G. (1992) Milk fat. Gas chromatographic analysis of triglycerides in assessment of authenticity of butter. *Scienza Tecnica Lattiero Casearia*, **43**, 95–110.

Lewkowitsch, J. (1904) *Chemical Technology and Analysis of Oils, Fats and Waxes*, 3rd edn. Macmillan, London, p. 364.

Lipp, M. (1995) Review of methods for the analysis of triglycerides in milk fat: application for studies of milk quality and adulteration. *Food Chem.*, **54**, 213–221.

Lipp, M. (1996a) Determination of adulteration of butter fat by its triglyceride composition obtained by GC. A comparison of the suitability of PLS and neural networks. *Food Chem.*, **55**, 389–395.

Lipp, M. (1996b) Comparison of PLS, PCR, and MLR for the quantitative determination of foreign oils and fats in butter fats of several European countries by their triglyceride composition. *Z. Lebensm. Unters. Forsch.*, **202**, 193–198.

Lotito, A. and Cucurachi, A. (1966) About the acidic composition of lard. *Riv. Ital. Sost. Grasse*, **43**, 499–504.

Luf, W. (1987) Detection of foreign fat in milk fat by triglyceride analysis. 1. Seasonal and regional variations in the triglyceride composition of Austrian milk fat. *Oesterreichische Milchwirt*, **42**, 23–28.

Luf, W. (1988a) Detection of adulteration of milk fat. 1. *Deutsch Milchwirtsch.*, **39**, 699–701, 735–737.

Luf, W. (1988b) Detection of adulteration of milk fat. 2. *Deutsch Milchwirtsch.*, **39**, 735–737.

Luf, W., Stock, A. and Brandl, E. (1987) Detection of foreign fat in milk fat by triglyceride analysis. 2. Detecting dulteration by evaluating triglyceride combinations. *Oesterreichische Milchwirt*, **42**, 29–35.

Maniongui, C., Gresti, J., Bugaut, M., Gauthier, S. and Bezard, J. (1991) Determination of bovine butterfat triacylglycerols by reversed-phase liquid chromatography and gas chromatography. *J. Chrom.*, **543**, 81–103.

Mariani, C., Venturini, S., Fedeli, E. and Contarini, G. (1994) Detection of refined animal and vegetable fats in adulteration of pure milkfat. *J. Am. Oil Chem. Soc.*, **71**, 1381–1384.

Maraini, C. and Venturini, S. (1997) Characterization of desterolised high oleic sunflower oil in olive oil. II. *Riv. Ital. Sost. Grasse.*, **74**, 489–500.

Maritano de Correche, M. and Oxley, R. (1985) Total fatty acid, aliphatic alcohol, and hydrocarbon compositions in edible fats. *Grasas Aceitas*, **36**, 88–92.

Mathew, T.V. and Kamath, K.S. (1978) Detection of vegetable oils in ghee and *vice versa* by reverse phase thin layer chromatography. *Research and Industry, India*, **23**, 168–169.

Matter, L. (1988) Capillary gas-chromatographic detection of the adulteration of sheep cheese. *Lebensm. Ger. Chem.*, **42**, 8–9.

Matter, L. (1991) Identification of animal fats and mixtures thereof by capillary gas chromatography. *Fett. Wissen. Technol.*, **93**, 536–538.

Matter, L. (1992) Determination of the animal of origin of dairy products and raw and cooked meats by GC analysis of the fatty acid methyl esters (FAME) obtained by transesterification. *J. High Res. Chrom. Chrom. Commun.*, **15**, 514–516.

Min, H.L., Nan, W.S., Ming, H.Y., Mei, L.W. and Youk, M.C. (1998) A rapid method for direct determination of free cholesterol in lipids. *J. Chinese Agr. Chem. Soc.*, **36**, 123–133.

Montiel-Sosa, J.F., Ruis-Pesini, E., Montoya, J., Roncales, P., Lopez-Perez, M.J. and Perez-Martos, A. (2000) Direct and highly species-specific detection of pork meat and fat in meat products by PCR amplification of mitochondrial DNA. *J. Agr. Food Chem.*, **48**, 2829–2832.

Mottram, H.R. (1999) *The Application of HPLC-APCI MS to the Regiospecific Analysis of Triacylglycerols in Edible Oils and Fats.* PhD thesis, Department of Chemistry, University of Bristol, UK.

Movia, E. and Remoli, S. (1977) Application of enzymic hydrolysis to determine the genuineness of butter., *Bollettino dei Chimici dei Laboratori Provinciali*, **3**, 187–192.

Murphy, J.J., Connolly, J.F. and McNeill, G.P. (1995) Effects on cow performance and milk fat composition of feeding full fat soyabeans and rapeseeds to dairy cows at pasture. *Livestock Prod. Sci.*, **44**, 13–25.

Najera, A.I., Perea, S., Barcina, Y., de Renobales, M. and Barron, L.J.R. (1999) A reverse-phase high performance liquid chromatographic method for analyzing complex mixtures of triglycerides. Application to the fat fraction of an aged cheese. *J. Am. Oil Chem. Soc.*, **76**, 399–407.

Nakazawa, Y. (1981) The distribution of sterols in mixtures of oils and fats, in *Collected essays by members of the Faculty, Kyoritsu Women's Junior college (Department of Domestic Science)*, No. 24, 83–90.

Offer, N.W., Marsden, M., Dixon, J., Speake, B.K. and Thacker, F.E. (1999) Effect of dietary fat supplements on levels of n-3 polyunsaturated fatty acids, trans acids and conjugated linolenic acid in bovine milk. *Animal Sci.*, **69**, 613–625.

Parodi, P.W. (1969) Detection of synthetic and adulterated butterfat. 2. Trisaturated triglyceride analysis. *Aus. J. Dairy Technol.*, **24**, 190–195.

Parodi, P.W. (1973a) The sterol content of milkfat, animal fats margarines and vegetable oils. *Aus. J. Dairy Technol.*, **28**, 135–137.

Parodi, P.W. (1973b) Detection of synthetic and adulterated butterfat. *Aus. J. Dairy Technol.*, **28**, 38–42.

Parodi, P.W. and Dunstan, R.J. (1969) Detection of synthetic and adulterated butterfat. II. Trisaturated triglyceride analysis. *Aust. J. Dairy Technol.*, **24**, 190–195.

Perea, S., de Labastida, E.F., Najera, A.I., Chavarri, F., Virto, M., de Renobales, M. and Barron, L.J.R. (2000) Seasonal changes in the fat composition of lacha sheep's milk used for Idiazabal cheese manufacture. *Eur. Food Res. Technol.*, **210**, 318–323.

Perrin, J.L. and Prevot, A. (1986) Use of a laser nephelometric detector in the study of fats by HPLC. 2. Analysis of triglycerides of oils and fats. *Rev. Franc. Corps Gras*, **33**, 437–445.

Peters, H. and Wieske, Th. (1966) Detection of traces of polybranched fatty acids in fats. *Fette Seif. Anstrichm.*, **68**, 947–950.

Podlaha, O. and Toregard, B. (1984) HPLC separation of mono-unsaturated triglycerides with respect to α- or β-position of oleic acid in some vegetable fats. *Fette Seif. Anstrichm.*, **86**, 243–245.

Polidori, P., Chiesa, L., Moretti, V.M. and Valfre, F. (1996) Milk fat quality: nutritional and sensorial parameters related to butter production. *Ind. Aliment.*, **35**, 8–12.

Povolo, M., Bonfitto, E., Contarini, G. and Toppino, P.M. (1999) Study on the performance of three different capillary gas chromatographic analyses in the evaluation of milk fat purity. *J. High Resol. Chrom. Chrom. Commun.*, **22**, 97–102.

Prager, M.J. (1989) Differential characteristics of fatty acids in cheese from milk of various animal species by capillary gas chromatography. *J. Assoc. Off. Anal. Chem.*, **72**, 418–421.

Precht, D. (1991) Detection of adulterated milk fat by fatty acid and triglyceride analysis. *Fett Wissen. Technol.*, **93**, 538–544.

Precht, D. (1992a) Detection of foreign fat in milk fat. 1. Qualitative detection by triglyceride formulae. *Z. Lebensm. Unters. Forsch.*, **194**, 1–8.

Precht, D. (1992b) Detection of foreign fat in milk fat. 2. Quantitative evaluation of foreign fat mixtures. *Z. Lebensm. Unters. Forsch.*, **194**, 107–114.

Precht, D. and Heine, K. (1986) Detection of modified milk fat by triglyceride analysis. 2. Detection of foreign fat in milk fat by triglyceride combinations. *Milchwissenschaft*, **41**, 406–410.

Precht, D., Molkentin, J. and de Froidmont-Goertz, I. (1998) Anhydrous butterfat reference material CRM 519: certification of triglyceride composition and cholesterol content including homogeneity and stability tests. *Fett. Lipid.*, **100**, 546–554.

Ramos, M., Martinez-Castro, I. and Juarez, M. (1977) Detection of cows milk in Manchego cheese. *J. Dairy Sci.*, **60**, 870–877.

Rashwan, M.R.A. and Youssef, M.K.E. (1989) Detection and evaluation of lard adulteration in pure goat and mutton tallow. *Proceedings of 35th International Congress of Meat Science and Technology*, **2**, 542–548.

Robinson, N.P. and Macgibbon, A.K.H. (1998) The composition of New Zealand milk fat triacylglycerols by reversed phase high performance liquid chromatography. *J. Am. Oil Chem. Soc.*, **75**, 993–999.

Roderbourg, H. and Kuzdzal-Savoie, S. (1979) The hydrocarbons of anhydrous butterfat: influence of technological treatments. *J. Am. Oil Chem. Soc.*, **56**, 485–488.

Rossell, B. (2001) Origins and Chemical Properties, in *Oils and Fats, Vol. 2., Animal Carcass Fats* (ed. B. Rossell), Leatherhead Publishing, Leatherhead, UK.

Ruiz-Sala, P., Hierro, M.T.G., Matrinez-Castro, I. and Santa-Maria, G. (1996) Triglyceride composition of ewe, cow and goat milk fat. *J. Am. Oil Chem. Soc.*, **73**, 283–293.

Saeed, T., Abu-Dagga, F. and Rahman, H.A. (1986) Detection of pork and lard as adulterants in beef and mutton mixtures. *J. Assoc. Off. Anal. Chem.*, **69**, 999–1002.

Saeed, T., Ali, S.G., Abdul-Rahman, H.A. and Sawaya, W.N. (1989) Detection of pork and lard as adulterants in processed meat: liquid chromatographic analysis of derivatized triglycerides. *J. Assoc. Off. Anal. Chem.*, **72**, 921–925.

Sato, T. (1997) Application of near infra-red spectroscopy for the analysis of fatty acid composition. *Lipid Technol.*, **9**, 46–49.

Schneller, R. and Wullschleger, R. (1992) Quantitative determination of milk fat in fat mixtures by gas chromatographic analysis of triglycerides. *Mitt. Gebiete Lebensm. Hyg.*, **83**, 737–744.

Schulte, E. and Weber, N. (1991) Disteryl ethers in bleached oils and fats—comparison with the conjugated trienes. *Fett. Wissen. Technol.*, **93**, 517–518.

Seher, A. (1987) The cholesterol content of plant oils. *Fat Sci. Technol.*, **89**, 27–30.

Shi, Y., Smith, C.M. and Hartel, R.W. (2001) Compositional effects on milk fat crystallization. *J. Dairy Sci.*, **84**, 2392–2401.

Sieber, R., Badertscher, R., Buetikofer, U., Collomb, M., Eyer, H. and Nick, B. (1998) Composition of Swiss butter. *Mitt. Gebiete Lebensm. Hyg.*, **89**, 84–86.

Sol-Morales, M., Palmquist, D.L. and Weiss, W.P. (2000) Milk fat composition of Holstein and Jersey cows with control or depleted copper status and fed whole soybeans or tallow. *J. Dairy Sci.*, **83**, 2112–2119.

Soliman, M.A. and Younes, N.A. (1986) Adulterated butter fat: fatty acid composition of triglycerides and 2-monoglycerides. *J. Am. Oil Chem. Soc.*, **63**, 248–250.

Spanos, G.A., Schwartz, S.J., van Breemen, R.B. and Huang, C.-H. (1995) High performance liquid chromatography with light scattering detection and desorption chemical-ionization tandem mass spectrometry of milk fat triacylglycerols. *Lipids*, **30**, 85–90.

Sue, J.L., Mao, C.C. and Chin, W.L. (2001) Detection of cow's milk in goats milk by capillary zone electrophoresis. *Aus. J. Dairy Technol.*, **56**, 24–27.

Thorpe, C.W. (1970) Comparison of separation processes for identification of oils by gas chromatography. *J. Assoc. Off. Anal. Chem.*, **53**, 623–628.

Tiscornia, E., Camurati, F., Gastaldo, P. and Pagano, M.A. (1976) The sterolic fraction of tomato seed oil (*Lycopersicon esculentum* Miller). *Riv. Ital. Sost. Grasse*, **53**, 119–129.

Toppino, P.M., Contarini, G., Traversi, A.L. Amelotti, G. and Gargano, A. (1982) Gas chromatographic parameters of genuine butter. *Riv. Ital. Sost. Grasse*, **59**, 592–610.

Tunick, M.H., Malin, E.L., Shieh, J.J., Smith, P.W. and Holsinger, V.H. (1998) Detection of mislabelled butter and cheese by differential scanning calorimetry and rheology. *Am. Lab.*, **30**, 28, 30–32.

Ulberth, F. (1994) Detection of milk fat adulteration by linear discriminant analysis of fatty acid data. *J. Assoc. Off. Anal. Chem.*, **77**, 1326–1334.

Ulberth, F. (1995) Quantitation of foreign fat in foreign fat/milkfat mixtures by multivariate regression analysis of fatty acid data. *J. Agr. Food Chem.*, **43**, 1556–1560.

Ulberth, F. and Buchgraber, M. (2000) Authenticity of oils and fats. *Eur. J. Lipid Technol.*, **102**, 687–694.

Uotila, M. (1996) Goat and sheep milk cheeses: detection of adulterations. *Thesis, University of Helsinki*, Helsinki, Finland.

Valentinis, G. (1970) Study in abnormal analytical variations of mountain butters. *Ind. Aliment.*, **9**, 103–107.

Vallejo-Cordoba, B. (1999) Detection of milk fat substitution by applying multivariate analysis to fatty acid derivatives prepared in situ, in *Proceedings of Annual Meeting of Institute of Food Technology, 1999* (http://confex.com/ift/99annual/abstracts/3794.htm).

Vanoni, M.C., Colombini, M. and Amelotti, G. (1978) Adulteration of butter with beef suet and lard. 1. GLC analysis of fatty acids. *Riv. Ital. Sost. Grasse*, **55**, 320–325.

Verbeke, R. and de Brabander, H. (1979) An alternative method for the detection of pork fat adulteration with beef tallow, in *Proceedings of the 25th European Meeting of Meat Research Workers*, 767–772.

Verbeke, R. and de Brabander, H. (1980) Identification of animal fat species, in *Proceedings of the 26th European Meeting of Meat Research Workers*, **1**, 150–153.

Wrezel, P.W., Krishnamurphy, R.G. and Hasenhuettl, G.L. (1992) Method for removing cholesterol from edible oils, U.S. Patent 5128162 (Kraft General Foods).

Wurziger, J. and Hensel, G. (1969) Determination of the nature of meat from the fat components and substances accompanying fats. *Fette Seif. Anstrichm.*, **71**, 144–151.

Yen, G.C. and Chen, C.J. (2000) Effects of fractionation and the refining process of lard on cholesterol removal by beta-cyclodextrin. *J. Food Sci.*, **65**, 622–624.

Youk, M.C., Mei, L.W., Kow, L.K. and Min, H.L. (1995) Removal of cholesterol from lard by liquid–liquid extraction. *J. Chinese Agr. Chem. Soc.*, **33**, 94–103.

Youk, M.C, Hsiu, J.L., Chung, W.C. and Mei, L.W., (1999) A rapid gas chromatographic method for direct determination of free sterols in animal and vegetable fats. *Yaowu Shipin Fenxi (J. Food Drug Anal.)*, **7**, 279–290.

Younes, N.A. and Soliman, M.A. (1987) Adulteration of butterfat: sterol composition. *Grasasa Aceitas*, **38**, 372–374.

Youssef, M.K.E. and Rashwan, M.R.A. (1987) Detection and evaluation of lard adulteration in pure buffalo and cow butter, in *Proceedings of the 33rd European Meeting of Meat Research Workers*, **2**, 374–376.

Youssef, M.K.E. and Rashwan, M.R.A. (1989) GLC Detection and assessment of lard adulteration in buffalo tallow, in *Proceedings of the 35th International Congress of Meat Science and Technology*, **2**, 549–555.

6 Analysis of minor components as an aid to authentication

Michael H. Gordon

6.1 Introduction

Minor components in crude edible oils may include sterols and related compounds, tocopherols, hydrocarbons, pigments, lignans and phenols, together with a range of volatile components. Most edible oils are refined before use and the refining process is very effective at removing many of the minor components, including volatile components, steryl glycosides, pigments, some lignans and phenols. Consequently, in many cases the authentication of refined oils involves analysis of the bulk components—especially triacylglycerols—although analysis of sterols, tocopherols and lignans can be useful for selected oils.

6.2 Sterols and related compounds

6.2.1 Sterols

Sterols constitute the majority of the unsaponifiable matter in edible oils and account for up to 1% of the oil in most cases, although contents in excess of 2% may be found in some oils (Table 6.1). Sterols mainly occur as mixtures of free sterols and steryl esters. However, glycosides occur as minor components in crude vegetable oils, although they are removed during refining. In glycosides, the 3-hydroxyl position of the sterol is linked to position 1 of a sugar, and the 6-position of the sugar can also be esterified to a fatty acid to form an acylated steryl glycoside. The predominant sterol in animal fats is cholesterol but vegetable oils contain mixtures of phytosterols, which include campesterol, stigmasterol, β-sitosterol, brassicasterol and Δ^5-avenasterol (Figure 6.1). Sterols differ mainly in the nature of the side chain, but some sterols, e.g. ergosterol, also differ in the double bond structure of the B ring. The mass ratio of free sterols:steryl esters in vegetable oils varies quite widely, with values of 0.74, 0.3–0.88, 1.1, 2.36, 0.41 being calculable from literature data for sunflower oil, rapeseed oil, sesame oil, soyabean oil and corn oil, respectively (Johansson, 1979; Johansson and Appelqvist, 1978; Kamal-Eldin and Appelqvist, 1994; Ferrari *et al.*, 1996). The sterol composition may show marked differences between the free sterol and steryl ester fractions of an oil (Kamal-Eldin and Appelqvist, 1994). The steryl ester fraction is commonly saponified before separate analysis of the sterols

Table 6.1 Desmethylsterol content of vegetable oils (data from *Codex Alimentarius* Commission, 1993, and other sources)

Oil/fat	Cholesterol	Brassicasterol	Campesterol	Stigmasterol	Sitosterol	Δ^5-avenasterol	Δ^7-stigmastenol	Δ^7-avenasterol	Others	Total (mg/kg)
Cocoa butter	1		8.0–11.0	24–31	58–63	3.0–5.0	1			
Coconut	0.6–3	<0.9	7.5–11	11.0–16	32–51	20–41	<3	<3	<4	470–1140
Corn (maize)	0.2–0.6	<0.2	18–24	4–8.0	54–67	4–8.0	1–4.0	0.7–3	<2	7950–22150
Cottonseed	0.7–2.3	<0.9	6.0–15	2.0–7.0	76–87	1.8–7.3	<1.4	0.8–3	<1.5	2690–6430
Hazelnut			5	1	93	2				1200
Linseed	<2		26–29	7–10.0	46–57	9.0–13	1.0–4.0			
Olive	<0.5	<0.1	<4	<4	>75	4.0–14	<0.5	<0.1		>1000
Palm	2.6–6.7		18–28	8.0–14	50–62	0–3	0.2–2	<5		370–1630
Palm kernel	0.6–3.7	<0.8	8.0–13	12.0–17	62–73	1.0–9.0	<2	<0.4	<3	790–1410
Peanut (groundnut)	<4	<0.2	12.0–20	5.0–13	47–68	8.0–19	<5	<6	<1.4	900–2860
Rapeseed		12.0–13.0	30–33	0.4–0.6	49–55	1.0–2.0				3500–6000
Rapeseed, (low erucic)	<1.3	5.0–13	25–39	<1	45–58	3.0–7.0	<1	<1	<4	4820–11280
Rice bran			20–28	8.0–15	49–54	5.0–11.0	1.0–2.0	2.0–4.0		10,550
Safflower	<0.5		9.0–13.0	6.0–10.0	40–50	2.0–4.0	16–23	3.0–5.0	0.5–3.0	2090–2650
Safflower (high oleic)	<0.2		10.0–16.0	8.0–15.0	52–60	5.0–6.0	13–18	5.0–6.0		
Sesame	<0.2	0.1–0.2	10.0–20	3.0–7	57–62	6.0–8	2.0–8	1.0–6	0.7–9	4500–18,960
Shea nut			38					11	51	2470
Soyabean	0.6–1.4	<0.3	15–24	16–19	52–58	2.0–4	1.0–5	1.0–5	<2	1830–4090
Sunflower	<1.3	<0.2	7.0–13	8.0–11	56–63	<7	7.0–13	3.0–7	<5	2430–4550
Sunflower (high oleic)			8.0–10	8.0–10	55–58	<3	14–22	4.0–6		
Walnut			5		89	5				1760
Wheat germ			19–29	1.0–4	60–67	6	3	2		5530

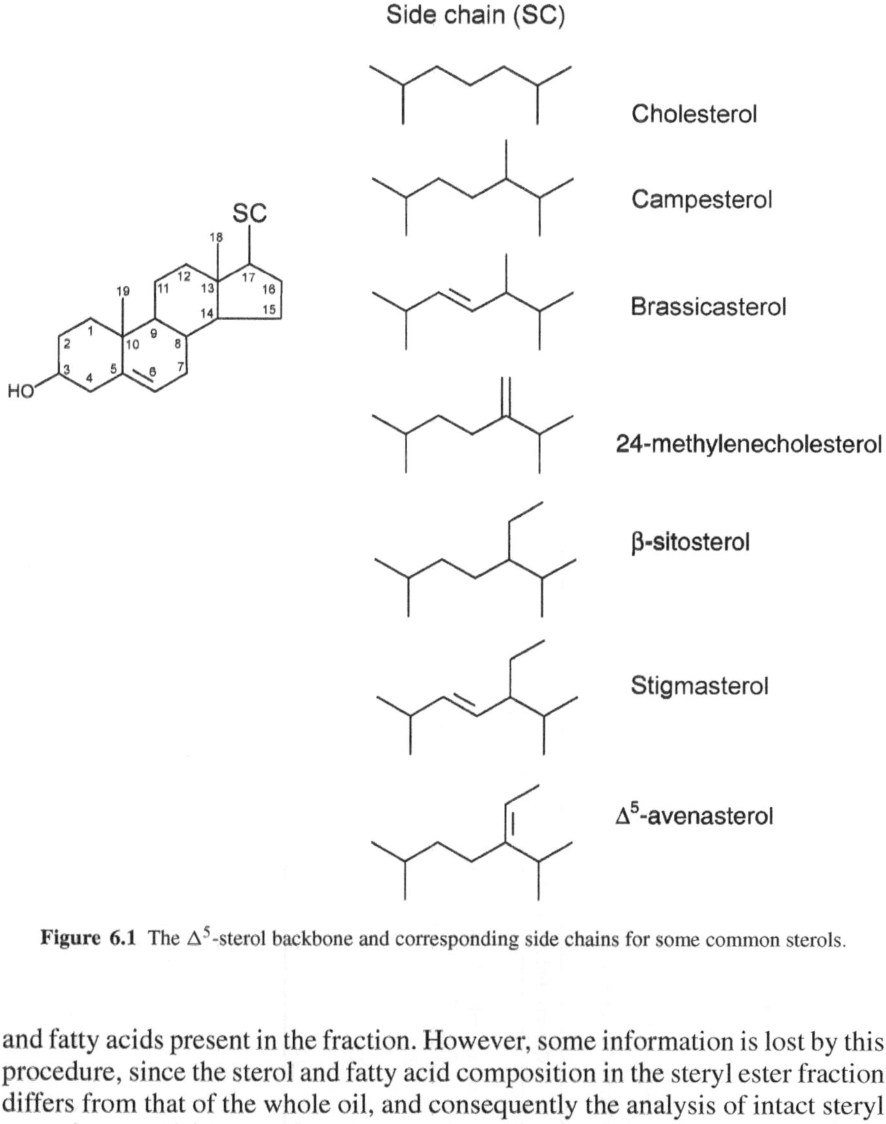

Figure 6.1 The Δ^5-sterol backbone and corresponding side chains for some common sterols.

and fatty acids present in the fraction. However, some information is lost by this procedure, since the sterol and fatty acid composition in the steryl ester fraction differs from that of the whole oil, and consequently the analysis of intact steryl esters is a promising technique for detection of adulteration (Gordon and Miller, 1997).

The most common sterols are Δ^5-unsaturated desmethylsterols, but 4α-methylsterols and 4,4-dimethylsterols occur at lower concentrations in vegetable oils. For example, sesame seed oil sterols are 85–89% des-, 9–11% mono- and 2–4% dimethylsterols (Kamal-Eldin and Appelqvist, 1994). The composition of monomethylsterol and dimethylsterol fractions of some oils is given in Table 6.2.

Table 6.2 Composition of monomethylsterol and dimethylsterol fractions of vegetable oils (other components unidentified) (Johansson and Croon, 1981; Kamal-Eldin and Appelqvist, 1994)

	Monomethyl sterols (%)				Dimethyl sterols (%)				
Oil/Fat	Obtusifoliol	Gramisterol	Cycloeucalenol	Citrostadienol	β-Amyrin	α-Amyrin	Cycloartenol	24-methylene cycloartanol	Cyclobranol
Sesame	21–25	20–23	6–10	21–29	5–8	5–7	41–50	26–34	0
Corn	21	26	6	29	1	2	43	40	0
Olive	7	3	14	22	4	3	18	31	10
Rapeseed	26	21	17	16	0	3	49	37	0
Soyabean	6	9	10	44	9	5	26	13	trace
Sunflower	26	15	3	38	5	11	19	48	trace

Although many vegetable oils are quite similar in sterol composition, there are some important differences. Detection of brassicasterol can be used to detect adulteration of many oils by rapeseed oil as this sterol is only present in most oils in small amounts but rapeseed oil contains up to 780 mg/kg brassicasterol (see Table 6.1). The detection of adulteration of vegetable oils with animal fats can be achieved by analysis of the cholesterol content, since this sterol is either absent or present at very low concentrations in vegetable oils.

6.2.2 *Effect of refining on the sterol content of oil*

Each refining step has the effect of reducing the total sterols present. However, the relative proportions of individual sterols do not change significantly under normal refining conditions. The degree to which sterols are removed during refining depends on the refining conditions used. The bleaching step is responsible for the greatest reduction in the sterol content due to adsorption on the bleaching earth (Johansson and Hoffman, 1979). Gutfinger and Letan (1974) reported a reduction in sterol content from 3870 mg/kg to 3050 mg/kg during bleaching of soyabean oil. Ferrari *et al.* (1996) reported losses of 18–36% by the refining of corn, soyabean and rapeseed oils. Physical refining or the deodorization step in a chemical refining process reduces the content of free sterols in the oil, and a small proportion of the free sterols suffers dehydration leading to the formation of a conjugated diene from Δ^5-sterols. Losses of steryl esters are smaller than losses of free sterols during refining with reductions of 16%, 6% and 3% observed during the refining of corn, soyabean and rapeseed oil, respectively (Ferrari *et al.*, 1997).

In some cases where extreme refining conditions have been used, complete removal of the sterols, both free and esterified, may occur (Grob *et al.*, 1994). By removing the sterols from an oil in this way, it is possible to prepare an adulterant that is undetectable by sterol analysis. As with adulteration of virgin olive oil with refined olive oil, this type of adulteration may be detected by analysis of sterol degradation products.

6.2.3 *Analysis of sterols*

Standard methods of analysis of total sterol content of oils involve saponification of the oil, followed by extraction and isolation of total sterols from the unsaponifiable fraction by thin layer chromatography (TLC) (AOCS, 1998). Quantification of individual sterols involves silylation of the sterol fraction and analysis by gas chromatography (GC). Sterols and steryl esters in oils and fats can be analysed by LC–GC after silylation or acylation of the free sterols (Artho *et al.*, 1993). An alternative approach to the analysis of intact steryl esters involves separation of sterols and steryl esters by solid phase extraction, column chromatography or TLC prior to analysis of steryl esters by GC.

6.2.4 *Detection of adulteration of pressed oil by addition of refined oil based on steradiene analysis*

Pressed oils are high value products because of their unique flavour and the adulteration of pressed oils with cheaper refined oil is a potential problem. The adulteration of virgin olive oil with refined olive oil is a common problem. Detection of this type of fraud is more difficult than detection of admixtures of different oils because analysis of most components, including triglycerides, sterols and tocopherols, cannot reveal the adulteration. Other components, e.g. phenols and volatiles, may be removed by refining, but again this does not represent a basis for developing a method of detecting refined oil in a pressed sample of the same oil. It is possible to detect this type of adulteration by analysing sterol degradation products formed during the refining process.

In 1974, Homberg (1974) reported that the action of bleaching earth on cholesterol in solutions of both synthetic triacylglycerol (TAG) and hexane caused the formation of the dehydration product cholesta-3,5-diene. Subsequently it was reported that bleaching of butterfat resulted in the formation of cholesta-3,5-diene (Roderbourg and Kuzdzal-Savoie, 1979) and the authors proposed that the detection of this artefact could be used to identify refined butterfat. In a review article (Kochhar, 1983), it was reported that several other authors had identified steradienes in bleached vegetable oils and proposed that the detection of these could be used to identify refined oils or mixtures of refined and unrefined oil.

A method of detection based on analysis of sterol dehydration products is likely to be very effective, because it is difficult to envisage a simple method of completely removing steradienes from an oil once they are formed. Most work on the development of the method has been performed to detect refined olive oil added to virgin olive oil, but the method should be applicable for the detection of refined oils to any pressed oil. The analysis of steradienes has been proposed as a method of detecting refined vegetable oils in chocolate (Crews *et al.*, 1997). However, one drawback of this method is that it yields little information on the amount of oil added in a blend because the concentration of sterol degradation products can vary widely depending on the refining procedure and conditions used (Tsimidou, 1995).

In olive oil, the main sterol is β-sitosterol and this dehydrates to a mixture of 2,4- and 3,5-stigmastadienes on heating to high temperatures or in the presence of a bleaching earth (Grob and Bronz, 1994; Cert and Moreda, 1998).

Cert *et al.* (1994) reported that bleaching caused the formation of the highest levels of stigmasta-3,5-diene and that bleaching temperature and earth type had an effect on sterol dehydration. These authors also reported that heating olive oil at temperatures below 150°C for 15 min did not cause stigmastadiene formation. However, heating for this time at 175°C and 200°C caused the formation of 0.025 and 0.076 mg/kg stigmastadiene, respectively.

The energy of activation for sitosterol dehydration was calculated as 191.0 ± 9.0 kJ/mol, but this is reduced to 70.8–80.6 kJ/mol in the presence of bleaching earths (Gordon and Firman, 2001). The rate of dehydration of other sterols is expected to be similar to that of sitosterol.

The earth that was most active for pigment removal was also most active for sitosterol dehydration. The stigmasta-3,5-diene concentration in 250 virgin olive oils has been analysed and was found to be negligible (Cert *et al.*, 1994). These authors also determined that neither storage nor oxidation of olive oil caused the formation of steradienes. Therefore it is reasonable to expect a virgin olive oil to be free of stigmasta-3,5-diene. However the legal limit allowed by the EU is 0.15 mg/kg to allow for small incidences of accidental adulteration of virgin olive oil by residues of refined oils adhering to oil extraction equipment such as filtering devices, centrifuges and common pipelines (Commission regulation (EC) No. 656/95).

Grob *et al.* (1992) reported the effect on stigmasta-3,5-diene formation of performing bleaching and deodorization sequentially on olive oil. They reported that bleaching formed large amounts of stigmasta-3,5-diene but that these were partially removed during deodorization.

Ferrari *et al.* (1996) claimed that deodorization caused significantly more steradiene formation than the other processes, including bleaching during the refining of corn, soyabean and rapeseed oils, but most other studies have shown that bleaching is the step at which most of the steradienes are formed.

The official method of analysis of steradienes in olive oil (Commission regulation (EC) No. 656/95) involves saponification of the oil with an internal standard of cholesta-3,5-diene, followed by extraction of the unsaponifiable fraction into hexane. The steradiene fraction is then separated from other hydrocarbons, such as alkanes and squalene isomers, by column chromatography on silica gel. Quantitative analysis is then performed by GC.

An alternative method involving coupled LC–GC has been proposed (Grob *et al.*, 1992), which considerably reduces analysis time as the only sample preparation is dilution of the oil to a 20% solution in hexane. HPLC removes the large amounts of triglycerides and isolates the steroidal hydrocarbons from other interfering components such as alkanes before online transfer of the steradiene fraction to GC.

A method using HPLC has also been proposed (Schulte, 1994) whereby an oil solution is first eluted through a silica gel column to separate the steradienes from the more polar lipids. The eluate is then concentrated and analysed on a reverse phase HPLC column using UV detection at 235 nm. This method involves a little more sample preparation than the LC–GC method proposed by Grob *et al.* (1992) but the equipment is more widely available.

6.2.5 *Formation of disteryl ethers*

Disteryl ethers are compounds consisting of two sterol groups linked at the 3-β carbon positions by an ether bridge. They are symmetrical compounds and, like steradienes, they are formed during refining but in lower concentrations (Weber *et al.*, 1992). Detection of disteryl ethers is further proof of refining but due to their lower abundance they are not as useful as steradienes as markers for refined oil.

6.3 Tocopherols and tocotrienols

Tocopherols and tocotrienols are absent from animal fats but they are normally present at up to 3500 mg/kg (0.35%) in crude vegetable oils, and similar amounts are present in refined oils if the refining procedure is performed carefully. Losses can occur during deodorization or physical refining, especially if the vacuum is not sufficiently high. Tocopherols and tocotrienols occur as mixtures of four homologues with saturated sidechains (α-, β-, γ- and δ-tocopherols), and four homologues with unsaturated sidechains (α-, β-, γ- and δ-tocotrienols) (Figure 6.2).

Table 6.3 Tocopherol (T) and tocotrienol (T3) content of vegetable oils (data from *Codex Alimentarius* Commission, 1993; Dionisi *et al.*, 1995, and other sources)

Oil/fat	α-T	β-T	γ-T	δ-T	α-T3	β-T3	γ-T3	δ-T3	Total
Cocoa butter	1.0–19	<10	18–196	<17					25–220
Coconut	<17	<11	<14		<44		<1		<44
Corn (maize)	23–573	<356	268–2468	23–75	<239		<450	<20	331–3716
Cottonseed	136–674	<29	138–746	<21					389–1185
Hazelnut	12–49	1–13		<1					33–58
Linseed	5		430–575	4.0–8					430–588
Olive	63–135		7.0–15						70–150
Palm	4–193	<234	<526	<123	4–336		14–710	<377	141–1465
Palm kernel	<44	<248	<257				<60		<257
Peanut (groundnut)	49–373	<41	88–389	<22					176–1291
Rapeseed	116	34	737	275					1165
Rapeseed (low erucic)	100–386	<140	189–753	<22					438–2680
Rice bran	576		324						900
Safflower	234–660	<17	<12				<12		246–664
Sesame	<3	<36	521–983	4.0–21			<20		531–1003
Soyabean	9–352	<45	89–2307	154–932					601–3363
Sunflower	403–935		<34	<7					447–1514
Sunflower (high oleic)	94–96	3.0–4	<2						98–102
Walnut	10.0–20		263–400	46–60					309–455
Wheat germ	880–1800	250–1150	175–700	<100	<200	<200			1400–2400

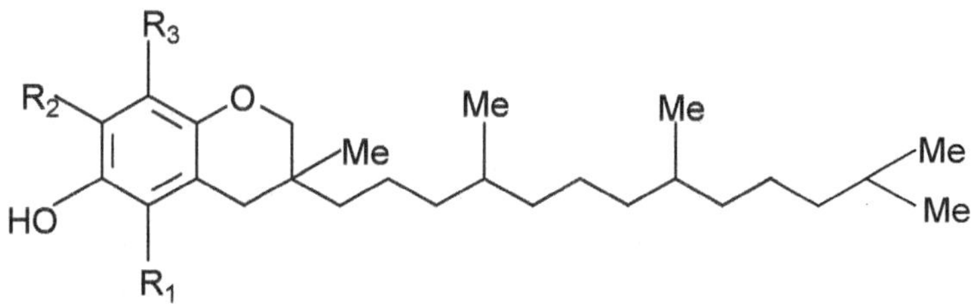

Name	R_1	R_2	R_3
α-tocopherol	Me	Me	Me
β-tocopherol	Me	H	Me
γ-tocopherol	H	Me	Me
δ-tocopherol	H	H	Me

Figure 6.2 The structure of tocopherols.

In some cases, analysis of tocopherols can be used to detect adulteration of oil. Both palm oil and grapeseed oil are relatively unusual in containing significant levels of tocotrienols, which are absent or present at much lower levels in most vegetable oils (Table 6.3). Detection of these compounds can be used to identify adulteration in olive oil at levels as low as 2% (Dionisi *et al.*, 1995).

The limitation of tocopherol analysis is that the tocopherol composition is similar for many oils. For many oils, including sunflower and olive oils, α-tocopherol is the main tocopherol present, but for some oils, e.g. soyabean oil, relatively high levels of γ- and δ-tocopherols may be present so some examples of adulteration may be detected on this basis. The tocopherol composition or high levels of tocopherols may help to indicate adulteration when considered with other analytical data. One problem with detecting adulteration based on tocopherol analysis is that tocopherols are reduced to variable extents by refining, depending on the refining conditions employed. Major losses of tocopherols occur at high temperatures in the presence of air.

Tocopherols can be analysed readily in vegetable oils by normal phase HPLC with UV or fluorescence detection. A standard method is described by AOCS (1998).

6.4 Fatty alcohols

Long chain alcohols, known as fatty alcohols, are present only in very small amounts in vegetable oil but are found in greater quantities in some marine oils (Sonntag, 1979). Fatty alcohols also exist in small amounts in vegetable oils as waxes, where they are esterified to a fatty acid. The wax component of sunflower, soyabean and peanut oil includes esters of long chain saturated fatty acids and a monounsaturated alcohol, mainly eicosenoic alcohol (Henon *et al.*, 2001). Waxes of this type are absent from corn and rice bran oil.

The wax ester content of olive oil varies widely in different grades of olive oil, being low in extra virgin or virgin olive oil, but it is much higher in solvent-extracted grades of olive oil and this has been accepted as a method of detecting adulteration of pressed olive oil with solvent-extracted olive oil (Nota *et al.*, 1999).

6.5 Phenols, lignans, secoiridoids and flavonoids

Pressed oils contain a range of phenols including flavonoids. The phenolic compounds found in virgin olive oils are simple phenols (hydroxytyrosol, tyrosol), phenolic acids (hydroxycinnamic and hydroxybenzoic acids), secoiridoids including oleuropein, tyrosol and hydroxytyrosol derivatives, and lignans including 1-acetoxypinoresinol and pinoresinol in virgin olive oil (Brenes *et al.*, 2000; Montedoro *et al.*, 1993; Owen *et al.*, 2000a,b; Tsimidou *et al.*, 1996). Phenolic compounds are removed in the neutralization step of a refining procedure. The removal of gossypol from crude cottonseed oil is essential, since it is known to cause male infertility (Hofmann, 1989). The lignans, sesamin and sesamolin are significant components in the unsaponifiable fraction of crude sesame seed oil, with typical levels of occurrence being 0.55% and 0.5%, respectively (Kamal-Eldin and Appelqvist, 1994). Sesamolin is converted into sesamol during industrial refining and during frying and converted into sesaminol during bleaching (Fukuda *et al.*, 1986a,b).

6.6 Hydrocarbons

Most oils contain low levels of saturated and unsaturated hydrocarbons. In olive oil, the unsaturated hydrocarbon squalene can constitute up to 40% of the unsaponifiable fraction (Boskou, 1996). Other hydrocarbons commonly present in olive oil are straight chain alkanes and alkenes with 13 to 35 carbon atoms, along with very low amounts of branched chain hydrocarbons. Variations are found between different olive varieties but the main hydrocarbons are those with 23, 25, 27 and 29 carbon atoms (Guinda *et al.*, 1996). Olive oil can clearly be differentiated from other vegetable oils on the basis of hydrocarbon components, and levels of 2.6% crude rapeseed oil or crude sunflower oil can be detected by hydrocarbon analysis (Webster *et al.*, 1999). Terpenes have been identified in the volatile fraction of crude sunflower oil (Bocci and Frega, 1996).

6.7 Other components

Apart from the compounds mentioned above, crude vegetable oils also contain volatile compounds that may give some oils a characteristic aroma. The presence of (*E*)-5-methylhept-2-en-4-one gives pressed hazelnut oil a characteristic aroma, and the detection of this component by GC–MS has been developed into a sensitive method for the detection of this oil in virgin olive oil (Blanch *et al.*, 2000). Detection of hazelnut oil in olive oil is difficult by analysis of TAGs, fatty acids, sterols or tocopherols because the differences in composition between the oils are small relative to the natural variability of composition of hazelnut and olive cultivars. Recent work has also shown differences in the minor non-volatile components that can be extracted from pressed hazelnut oil and virgin olive oil, and a component detectable in extracts from pressed hazelnut oil by HPLC–UV analysis is promising for the detection of this oil in a mixture with virgin olive oil (Gordon *et al.*, 2001).

6.8 Conclusion

The analysis of minor components is a useful method of detecting adulteration of some pressed oils. For many refined oils, analysis of sterols or tocopherols may be less likely than analysis of bulk components, especially TAGs, to be an effective method of detecting adulteration, but the minor component composition may be helpful when considered with other analytical data.

References

AOCS (American Oil Chemists' Society) (1998) *Official Methods and Recommended Practices of the AOCS*, 5th edition (ed. D. Firestone), AOCS, Champaign, IL, USA.

Appelqvist, L.A., Kornfeldt, A. and Wennerholm, J. (1981) Sterols and sterol esters in some *Brassica* and *Sinapis* seeds. *Phytochem.*, **20**, 207–210.

Artho, A., Grob, K. and Mariani, C. (1993) Online LC–GC for the analysis of the minor components in edible oils and fats. *Fett. Wissen. Technol.*, **95**(5), 176–180.

Blanch, G.P., del Mar Caja, M., Leon, M. and Herraiz, M. (2000) Determination of (*E*)-5-methylhept-2-en-4-one in deodorised hazelnut oil. Application to the detection of adulterated oils. *J. Sci. Food Agric.*, **80**, 140–144.

Bocci, F. and Frega, N. (1996) Analysis of the volatile fraction from sunflower oil extracted under pressure. *J. Amer. Oil Chem. Soc.*, **73**(6), 713–716.

Boskou, D. (1996) Olive oil quality, in *Olive Oil Chemistry and Technology* (ed. D. Boskou), AOCS Press, Champaign, IL, USA, pp. 101–103.

Brenes, M., Hidalgo, F.J., Garcia, A., Rios, J.J., Garcia, P., Zamora, R. and Garrido, A. (2000) Pinoresinol and 1-acetoxypinoresinol, two new phenolic compounds identified in olive oil. *J. Amer. Oil Chem. Soc.*, **77**(7), 715–720.

Cert, A. and Moreda, W. (1998) New method of stationary phase preparation for silver ion column chromatography: application to the isolation of steroidal hydrocarbons in vegetable oils. *J. Chromatogr.*, **823**, 291–297.

Cert, A., Lanzon, A., Carelli, A., Albi, T. and Amelotti, G. (1994) Formation of stigmasta-3,5-diene in vegetable oils. *Food Chem.*, **49**, 287–293.

Codex Alimentarius Commission (1993) Proposed draft standard for named vegetable oils, appendix 1 (CX 1993/16), Rome.

Commission regulation (EC) No. 656/95 of 28 March 1995 ammending regulation (EEC) No. 2568/91 on the characteristics of olive oil and olive residue oil and on the relevant methods of analysis and council regulation (EEC) No. 2568/87 on the tariff and statistical nomenclature and on the common customs tariff. *Off. J. Eur. Commun.*, **L69** (28 March 1995), 1–12.

Crews, C., Calvet-Sarrett, R. and Brereton, P. (1997) Analysis of sterol degradation products to detect vegetable fats in chocolate. *J. Am. Oil Chem. Soc.*, **74**(10), 1273–1280.

Dionisi, F., Prodolliet, J. and Tagliaferri, E. (1995) Assessment of olive oil adulteration by reverse-phase high performance liquid chromatography/amperometric detection of tocopherols and tocotrienols. *J. Am. Oil Chem. Soc.*, **72**(12), 1505–1511.

Ferrari, R., Schulte, E., Esteves, W., Bruhl, L. and Mukherjee, K. (1996) Minor constituents of vegetable oils during industrial processing. *J. Am. Oil Chem. Soc.*, **73**, 587–592.

Ferrari, R., Esteves, W. and Mukherjee, K. (1997) Alteration of steryl ester content and positional distribution of fatty acids in triacylglycerols by chemical and enzymatic interesterification of plant oils. *J. Am. Oil Chem. Soc.*, **74**(2), 93–96.

Fukuda, Y., Nagata, M., Osawa, T. and Namiki, M. (1986a) Contribution of lignan analogs to antioxidative activity of refined unroasted sesame seed oil. *J. Amer. Oil Chem. Soc.*, **63**(8), 1027–1033.

Fukuda, Y., Nagata, M., Osawa, T. and Namiki, M. (1986b) Acid transformation of sesamolin, the sesame oil constituent into an antioxidant bisepoxylignan, sesaminol. *Heterocycles*, **24**(4), 923–926.

Gordon, M.H. and Firman, C. (2001) Effects of heating and bleaching on formation of stigmastadienes in olive oil. *J. Sci. Food Agric.*, **81**(15), 1530–1532.

Gordon, M.H. and Miller, L.A.D. (1997) Development of steryl ester analysis for the detection of admixtures of vegetable oils. *J. Amer. Oil Chem. Soc.*, **74**, 505–510.

Gordon, M.H., Covell, C. and Kirsch, N. (2001) Detection of pressed hazelnut oil in admixtures with virgin olive oil by analysis of polar components. *J. Am. Oil Chem. Soc.*, **78**, 621–624.

Grob, K. and Bronz, M. (1994) Analytical problems in determining 3,5-stigmastadiene and campestadiene in edible oils. *Riv. Ital. Sost. Grasse*, **71**, 291–295.

Grob, K., Artho, A. and Mariani, C. (1992) Determination of raffination of edible oils and fats by olefinic degradation products of sterols and squalenes, using coupled LC–GC. *Fat Sci. Technol.*, **94**, 394–400.

Grob, K., Giuffre, A., Leuzzi, U. and Mincione, B. (1994) Recognition of adulterated oils by direct analysis of the minor components. *Fat Sci. Technol.*, **96**, 286–290.

Guinda, A., Lanzon, A. and Albi, T. (1996) Differences in hydrocarbons of virgin olive oils obtained from several olive varieties. *J. Agric. Food Chem.*, **44**, 1723–1726.

Gutfinger, T. and Letan, A. (1974) Quantitative changes in some unsaponifiable components of soya bean oil due to refining. *J. Sci. Food Agric.*, **25**, 1143–1147.

Henon, G., Recseg, K. and Kovari, K. (2001) Wax analysis of vegetable oils using liquid chromatography on a double absorbent layer of silica gel and silver nitrate-impregnated silica gel. *J. Amer. Oil Chem. Soc.*, **78**(4), 401–410.

Hofmann, G. (1989) *The Chemistry and Technology of Edible Oils and Fats and Their High Fat Products*. Academic Press, London.

Homberg, V. (1974) Alteration of sterols by industrial processing of fats and oils, I: Influence of refining conditions on the sterol content and sterol composition. *Fette Seif. Anstrichm.*, **76**(10), 433–435.

IUPAC (1987) Identification and determination of sterols by gas–liquid chromatography, in *Standard Methods of Analysis of Oils, Fats and Derivatives* (eds C. Paquot and A. Hautfenne), Oxford, Blackwell Scientific, pp. 165–169.

Johansson, A. (1979) The content and composition of sterols and sterol esters in sunflower and poppy seed oils. *Lipids*, **14**(3), 285–291.

Johansson, A. and Appelqvist, L.A. (1978) The content and composition of sterols and sterol esters in low erucic acid rapeseed. *Lipids*, **13**(10), 658–665.

Johansson, A. and Croon, L.B. (1981) 4-Demethyl-, 4-monomethyl- and 4,4-dimethylsterols in some vegetable oils. *Lipids*, **16**, 306–314.

Johansson, A. and Hoffman, I. (1979) The effect of processing on the content and composition of free sterols and sterol esters in soyabean oil. *J. Am. Oil Chem. Soc.*, **56**, 886–889.

Kamal-Eldin, A. and Appelqvist, L.A. (1994) Variations in the composition of sterols, tocopherols and lignans in seed oils from 4 sesamum species. *J. Amer. Oil Chem. Soc.*, **71**(2), 149–156.

Kochhar, S. (1983) Influence of processing on sterols of edible vegetable oils. *Prog. Lipid Res.*, **22**, 161–188.

Montedoro, G., Servili, M., Baldioli, M. and Miniati, E. (1993) Simple and hydrolyzable phenolic compounds in virgin olive oil. 3. Spectroscopic characterizations of secoiridoid derivatives. *J. Agric. Food Chem.*, **41**, 2228–2234.

Nota, G., Naviglio, D., Romano, R., Sabia, V., Musso, S. and Improta, C. (1999) Determination of the wax ester content in olive oils. Improvement in the method proposed by EEC regulation 183/93. *J. Agric. Food Chem.*, **47**, 202–205.

Owen, R.W., Mier, W., Giacosa, A., Hull, W.E., Spiegelhalder, B. and Bartsch, H. (2000a) Identification of lignans as major components in the phenolic fraction of olive oil. *Clin. Chem.*, **46**(7), 976–988.

Owen, R.W., Mier, W., Giacosa, A., Hull, W.E., Spiegelhalder, B. and Bartsch, H. (2000b) Phenolic compounds and squalene in olive oils. *Food Chem. Toxicol.*, **38**(8), 647–659.

Roderbourg, H. and Kuzdzal-Savoie, S. (1979) The hydrocarbons of anhydrous butterfat: influence of technological treatments. *J. Amer. Oil Chem. Soc.*, **56**, 485–488.

Schulte, E. (1994) Determination of edible fat refining of HPLC of Δ^5-steradienes. *Fat Sci. Technol.*, **96**, 124–128.

Sonntag, N. (1979) Structure and composition of fats and oils, in *Bailey's Oil and Fat Products* (ed. Daniel Swern), New York, John Wiley and Sons, pp. 1–99.

Tsimidou, M. (1995) The use of HPLC in the quality control of virgin olive oil. *Chromatogr. Anal.*, **38**, 5–7.

Tsimidou, M., Lytridou, M., Boskou, D., Pappa-Louisi, A., Kotsifaki, F. and Petrakis, C. (1996) On the determination of minor phenolic acids of virgin olive oil by RP–HPLC. *Grasas Aceites*, **47**, 151–157.

Weber, V., Bergenthal, D., Bruhl, L. and Schulte, E. (1992) Disteryl ethers—artifacts of bleaching of fats and oils. *Fat Sci. Technol.*, **94**, 182–192.

Webster, L., Simpson, P., Shanks, A.M. and Moffat, C.F. (1999) The authentication of olive oil on the basis of hydrocarbon concentration and composition. *Analyst*, **125**(1), 97–104.

7 Chemometrics as an aid in authentication

Ramón Aparicio and Ramón Aparicio-Ruiz

7.1 Introduction

Thousands of chemical compounds have been identified in oils and fats, although only a few hundred are used in authentication. This means that each object (food sample) may have a unique position in an abstract *n*-dimensional hyperspace. A concept that is difficult to interpret by analysts as a data matrix exceeding three features already poses a problem. The art of extracting chemically relevant information from data produced in chemical experiments by means of statistical and mathematical tools is called chemometrics. It is an indirect approach to the study of the effects of multivariate factors (or variables) and hidden patterns in complex sets of data. Chemometrics is routinely used for: (a) exploring patterns of association in data, and (b) preparing and using multivariate classification models. The arrival of chemometrics techniques has allowed the quantitative as well as qualitative analysis of multivariate data and, in consequence, it has allowed the analysis and modelling of many different types of experiments.

The general concept of authentication, on the other hand, is not only circumscribed to adulteration but also includes aspects such as characterization, geographical origin of foodstuffs, extraction and processing systems, among others. Thus, the combined effect of a great amount of data, supplied by the current sophisticated instrumentation and the various issues included inside the concept of authenticity means that the analyst faces a complex situation. We cannot therefore think that we only need to put data in one side of a 'black box' (the computer) and get the results from the other side.

7.2 Chemometric procedures in food authentication

The chemometric procedures that are currently applied in empirical investigations have been notably improved in recent years with the assistance of computer science. Researchers have passed from initial application of univariate analyses to extensive use of multivariate procedures in less than one decade. This qualitative step has been possible because, first, the new sophisticated analytical instruments are now able to analyse dozens of chemical compounds in hundreds of samples daily, and, second, due to personal computers that can work with a great diversity of software packages.

The application of mathematical algorithms has allowed conclusions to be arrived at that were unthinkable only a few decades ago. However, the conclusions will be useful only if analysts follow strictly the following three steps:

1. *Exploratory data analysis (EDA).* This analysis, also called 'pretreatment of data', is essential to avoid wrong or obvious conclusions. The EDA objective is to obtain the maximum useful information from each piece of chemico-physical data because the perception and experience of a researcher cannot be sufficient to single out all the significant information. This step comprises descriptive univariate statistical algorithms (e.g. mean, normality assumption, skewness, kurtosis, variance, coefficient of variation), detection of outliers, cleansing of data matrix, measures of the analytical method quality (e.g. precision, sensibility, robustness, uncertainty, traceability) (Eurachem, 1998) and the use of basic algorithms such as box-and-whisker, stem-and-leaf, etc.
2. *Bivariate statistics.* The objective here is to look for possible relationships between pairs of variables. Pearson's correlation has traditionally been the most used, although the analysis of the correlation matrix should be studied before the use of most multivariate statistical procedures.
3. *Multivariate statistics.* The main matter facing researchers in the authentication of foods characterized by numerous variables is how to organize the observed data into meaningful structures. Independent of the names given to the multivariate procedures, they can be clustered into two basic groups. The descriptive (non-supervised) group is characterized by the fact that there is no previous hypothesis classifying or defining the objects (samples). The procedures clustered inside this group analyse the information given and explore the data matrix in the search for new knowledge. There are many statistical procedures inside this group: factorial analysis (e.g. principal component analysis and maximum likelihood), canonical correlation, multiple partial correlation, clusters, correspondence analysis and multidimensional scaling and other less applied procedures. The explanatory (supervised) procedures, on the contrary, have the objective of checking *a priori* hypothesis or they are simply dependence models that can be subsumed under the general concept of regression. Discriminant analyses, multivariate analysis of variance, log-lineal models are, among many others, explanatory procedures also.

7.2.1 Pretreatment of data

Chemometrics and artificial intelligence procedures require a strict process of data selection. Since almost all statistical procedures are based on Bayesian theory, the information must comply with eight conditions:

1. The error of the method (or overall variability) and uncertainty should be known.
2. Chemical and/or physical analyses should be carried out in triplicate, or at least duplicate.

3. The assumption of normality should be verified with each variable.
4. Outliers should be detected and corrected, if possible.
5. The data set should have information on the whole range of values of the knowledge domain to be studied (e.g. the cultivars of a vegetable oil).
6. The validation (test) set should be independent of the training set, if possible.
7. The standardization of data should agree with the objective to be studied.
8. The number of variables (e.g. chemical compounds) should be lower than objects (e.g. samples).

Knowing which factors contribute to the overall variability it should be possible to improve the analytical methodology. The whole error (first condition) is composed of the systematic error (or bias), unspecified random errors, and a series of errors produced during chemical or physical analyses. Uncertainty, also expressed as standard deviation (type A uncertainty), is a concept for measuring the quality of the analytical procedures (Taylor and Kuyaat, 1994).

The great majority of statistical procedures are based on the assumption of normality of variables, and it is well known that the central limit theorem protects against failures of normality of the univariate algorithms. Univariate normality does not guarantee multivariate normality, though the latter is increased if all the variables have normal distributions; in any case, it avoids the deleterious consequences of skewness and outliers upon the robustness of many statistical procedures. Numerous transformations are also able to reduce skewness or the influence of outlying objects.

As chemometrics works almost exclusively with numbers, abnormal data can lead the experimentalist to obvious or wrong conclusions, e.g. outliers inside the training set of neural networks. Most of the outliers can be detected and some of them corrected before applying definitive mathematical procedures by so-called robust algorithms (Armstrong and Beck, 1990), although most statisticians usually remove them when the database is large enough. Erroneous conclusions are, however, mostly due to databases that do not keep all aspects of the food characterization (e.g. partial or skewed databases) or those that merge data of chemical compounds quantified with different techniques (e.g. fatty acids quantified by different chromatographic columns). A classical example is an authentication of the geographical origin of Italian virgin olive oil samples by artificial neural networks (ANN) (Zupan *et al.*, 1994). The differences between oils were mainly due to the use of different chromatographic columns (packed columns and open tubular columns versus capillary columns) when quantifying free fatty acids (FFAs) of oils from southern and northern Italy, respectively. The neural network has therefore thus mostly learned to recognize the different chromatographic columns.

Databases should be split into two independent parts: (a) the training set and (b) the test set. The first set is used to obtain the mathematical equations while the second set is used to validate the equations. The selection of objects (samples) for these sets should be carried out with random methods (e.g. random numbers) and the independent test set should comprise at least 25% of the total samples; samples of both sets must contain exemplars encompassing the appropriate variance over all relevant properties for the problem at hand. The use of an external validation set does not invalidate the use of any internal validation (e.g. cross-validation or leverage correction), which has been applied with success where the total number of objects was small. At this stage it is of great importance to have the correct exemplars in the training and test sets because if outliers are included in the construction of a model then inaccuracies in the predictions from new multivariate data are likely to occur.

Chemical or physical data can differ by orders of magnitude (e.g. ppb or ppm) and, moreover, data are collected from instruments that often give information in different scientific measurements (e.g. °C, % or g/l). In these cases, the scaling (also called standardization) should be applied in order to readjust the individual contributions to the outcome on an equal basis, so avoiding the problem that some variables can weight the results more than others. Furthermore, data standardization make residuals more symmetrically distributed, which is important because least square (LS) estimation is consistent with non-random residuals.

Overfitting is the commonest problem in multivariate statistical procedures when the number of variables is greater than objects (samples); one can 'fit an elephant' with enough variables. Tabachnick and Fidell (1983) have suggested minimum requirements for some multivariate procedures to avoid the overfitting or underfitting that can occur in a somewhat unpredictable manner, regardless of the multivariate procedure chosen.

7.3 Multivariate procedures

The main goal of this section is to provide a summary of several of the most widely used multivariate procedures in food authentication out of the vast array currently available. These are included in well-known computer packages such as BMDP, IMSL, MATLAB, NAG, SAS, SPSS and STATISTICA. The first three subsections describe unsupervised procedures, also called exploratory data analysis, that can reveal hidden patterns in complex data by reducing data to more interpretable information, to emphasize the natural grouping in the data and show which variables most strongly influence these patterns. The fourth and fifth subsections are focused on the supervised procedures of discriminant analysis and regression. The former produces good information when applied under the strictness of certain tests, whereas the latter is mainly used when the objective is calibration.

7.3.1 Cluster analysis

The term cluster analysis comprises classification algorithms designed to understand the information of data matrices, to describe the similarities and dissimilarities among objects (samples) and to single out categories grouping similar objects (Hartigan, 1975; Zupan, 1982). This collection encompasses the following algorithms: *K-means clustering*, which minimizes within-cluster variability while simultaneously maximizes between-cluster variability (Jacobsen and Gunderson, 1986); *block clustering*, which simultaneously amalgamates objects and variables; and *tree clustering* (also called joining clustering). The latter algorithm, which is the most popular (Massart and Kaufman, 1983), is based on two kinds of subprocedures: distance measures and amalgamation rules.

The hierarchical clustering method uses the distances (or dissimilarities) between variables when forming the clusters. The distances that can be computed are based on a single dimension or multiple dimensions:

(a) The Euclidean distance is the geometric distance in multidimensional space, and it is probably the most commonly chosen.
(b) The Squared Euclidean distance is similar to the preceding one though it adds progressively greater weight to objects that are further apart.
(c) Chebychev's distance is suggested to state that there are objects as 'different' on any one of the dimensions.
(d) The Power distance is applied when the analyst wants to increase, or decrease, the progressive weight on each dimension.
(e) The Manhattan (city block) distance is simply the average difference across dimensions. If this distance is applied, outliers should be previously removed or corrected because the effect of outliers is dampened with this distance.
(f) The Percent disagreement is a distance that is particularly useful if the information is categorical in nature.
(g) Mahalanobis' distance is a measure between two points in the space defined by two or more correlated variables. If the variables are correlated, the Mahalanobis distance will adequately account for the correlation whereas the simple Euclidean distance is not an appropriate measure. The results of Mahalanobis' distance will be identical to the Euclidean distance if the variables are uncorrelated.

Once several objects have been linked together, the next step is to determine the distances between the new clusters. This new procedure is carried out by linkage or amalgamation rules that determine when two clusters are similar enough as to be linked together. There are various possibilities:

(a) The K-nearest neighbour (single linkage) is determined by the distance of the two closest objects in the different clusters. Thus, the resulting clusters tend to represent long 'chains'.

(b) The K-furthest neighbour (complete linkage) is determined by the greatest distance between any two objects in the different clusters. This amalgamation method performs quite well with naturally distinct objects (or variables) but is inappropriate if the clusters tend to be elongated.
(c) The unweighted pair-group average is the average distance between all pairs of objects in two different clusters. This amalgamation method is not affected by the shape of clusters and, hence, it should be used when the objects form natural distinct groups.
(d) The weighted pair-group average should be used when the cluster sizes are suspected to be largely uneven. It is the same as the unweighted pair-group average method, except that the size of the respective clusters is used as a weight.
(e) The unweighted pair-group centroid is the centre of gravity for each cluster and the distance between two clusters is determined as the difference between centroids.
(f) The weighted pair-group centroid is appropriate when there are appreciable differences in cluster sizes. Although the same as the previous one, it takes into consideration differences in cluster sizes.
(g) Ward's method is different from all other methods because it uses the approach of the analysis of variance to evaluate the distances between clusters. This method is regarded as very efficient although it tends to create clusters of small size.

Clustering analysis has been applied to many authentication problems. Thus, the authentication of European monovarietal virgin olive oils has been analysed using various chemical compounds of unsaponifiable matter (Aparicio and Alonso, 1994) and volatile compounds (Aparicio *et al.*, 1997, 2000). The volatiles used for the example (Figure 7.1a,b) have not been explicitly selected to distinguish monovarietal oils but to point out possible differences between the chemometric procedures with respect to their results. The figures show how the election of the amalgamation rules and the linkage distances determine the result, so pointing out that analysts should meditate this subject before applying clustering analysis.

7.3.2 Factor analysis

The techniques for the modelling of complex data are clustered inside factor analysis (FA), principal components analysis (PCA) being the most applied in authentication. The objective of applying FA is to obtain a number of unobservable factors, from the original set of observable variables (e.g. chemical components or wave numbers), so as to reduce a large raw data matrix to a smaller one while retaining most of the original information. FA produces several linear combinations of observed variables generally called eigenvectors. The process

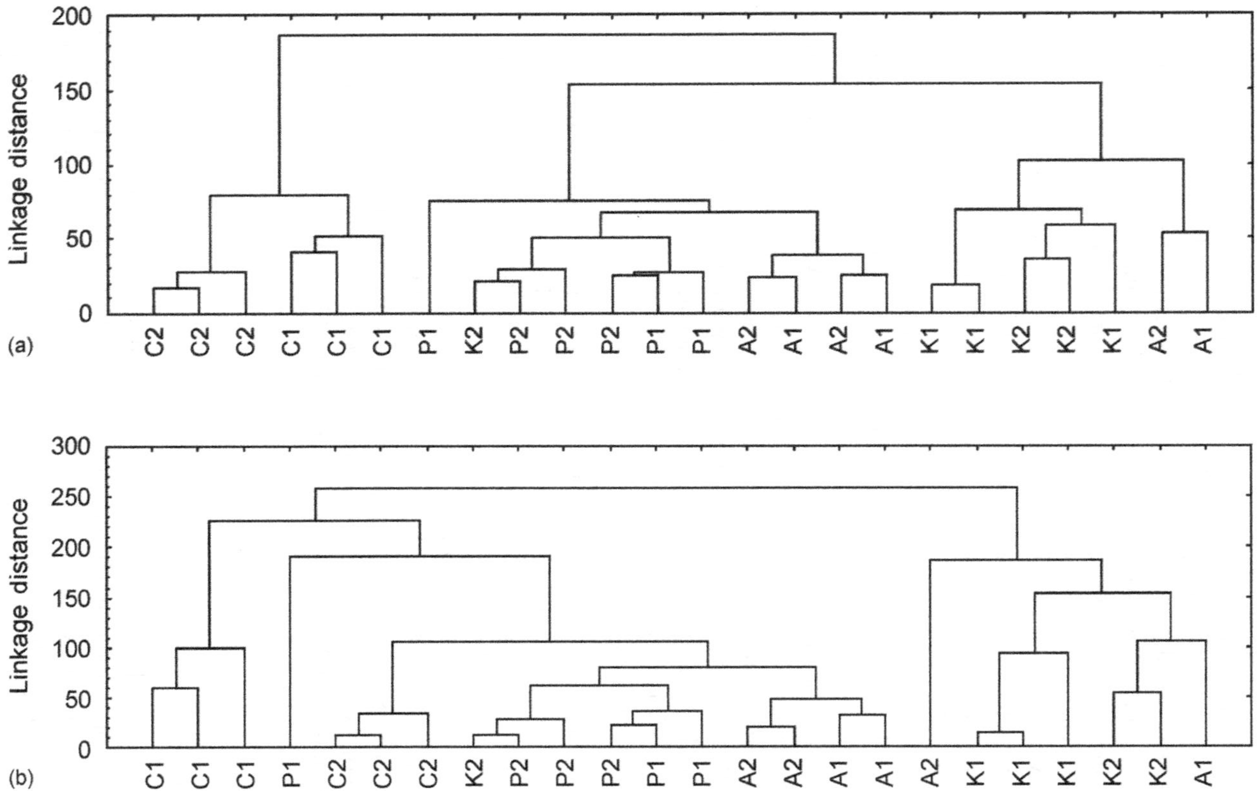

Figure 7.1 Authentication of monovarietal virgin olive oils: results of applying clustering analysis to volatile compounds. The Mahattan (city block) distance metric and Ward's amalgamation methods were used in (a) the Squared Euclidean distance and (b) complete linkage amalgamation methods. Note: A, cv. Arbequina (6); C, cv. Coratina (6); K, cv. Koroneiki (6); P, cv. Picual (6); 1, harvest 1991; 2, harvest 1992. Olives were harvested at three levels of maturity (unripe, normal, overripe) (source: SEXIA™ Group–Instituto de la Grasa, Seville, Spain).

of FA includes selecting a group of original variables, building the correlation matrix, determining the number of eigenvectors to be considered, extracting a set of eigenvectors from the correlation matrix, rotating the eigenvectors to increase interpretability and, eventually, making the conclusions. Conclusions should take into consideration that: (i) they must 'make sense'; (ii) they should be supported, when used in authentication, by the understanding from another science (chemistry, physics, biochemistry, agronomy, etc.) so demonstrating that results were not attained by chance.

The first step of FA is factor extraction; the methods described below are the most commonly used. The extraction methods calculate a set of orthogonal factors (or components) that in combination reproduce the matrix of correlation. The criteria used to generate the factors are not homogeneous for all methods but the differences between their solutions may be quite small.

(a) Principal components analysis (PCA) is the most used multivariate procedure as it is easy to interpret and permits an explanation for the maximum variability of initial distribution. Moreover, there is no need to invert a matrix when applying PCA.
(b) Multiple r^2 is used for estimating communalities. Prior to factoring, the diagonal of the correlation matrix (communalities) will be computed as the multiple r^2 of the respective variable with all other variables.
(c) The iterated communalities method uses multiple r^2 estimates for the communalities. The method adjusts the loadings over several iterations using the residual sums of squares to evaluate the goodness-of-fit of the resulting solution.
(d) The centroid method is a geometrical approach to FA (Wherry, 1984).
(e) The principal axis method first computes the eigenvalues from current communalities, and then the next communalities (sum of squared loadings) are repeatedly recomputed, based on the subsequent extracted eigenvalues and eigenvectors, to obtain minimum changes in communalities.
(f) Maximum likelihood factors calculate the loadings and communalities that maximize the likelihood of the correlation matrix. It needs an *a priori* hypothesis on the number of possible factors.

Although each of these methods of extracting factors has a different mathematical background, an algorithm that removed those eigenvectors that are dependent enough on the actual objects would improve the validity of the model. Cross-validation (Wold, 1978) is widely used to select eigenvectors that give the minimum residual sum of squares for the omitted objects (Piggott and Sharman, 1986). An additional check for the appropriateness of the extracted factors is to check deviations between reproduced and observed matrices, the new matrix being the matrix of residual correlations. This matrix may point to particular correlation coefficients that cannot be reproduced appropriately by the current number of factors.

Graphical representation is also important to visualize the results attained by the methods that extracted the factors. The objective is not to improve the fit between the observed and reproduced correlation matrices but as an aid to the interpretation of scientific results, making them more understandable. Eigenvector rotation is the most commonly used, and the four types of rotation are as follows:

(a) The Varimax normalized method (the most commonly used) performs a rotation of the normalized factor loadings; raw factor loadings divided by the square roots of the respective commonalities. This rotation maximizes the variances in the columns of the matrix of the squared normalized factor loadings.
(b) The Quartimax normalized method maximizes the variances in the rows of the matrix of the squared normalized factor loadings, but if the normalization has not been requested (Quartimax raw) the maximization is done in the squared raw factor loadings.
(c) Equamax rotation is a weighted mixture of Varimax and Quartimax rotations. It simultaneously maximizes the variances in the rows and columns of the matrix of the squared raw factor loadings.
(d) Oblique rotation has been developed to rotate factors without the constraint of orthogonality of factors, although they are often not easily interpreted (Wherry, 1984).

From an empirical point of view, three practical issues must be considered in the application of FA: (1) multicolinearity and singularity; (2) outliers among variables and with respect to the solution; and (3) validation of results. Extreme multicolinearity and singularity must be avoided for those algorithms that require matrix inversion. When multicolinearity is present (Tabachnick and Fidell, 1983) then it may be necessary to eliminate some variables. Variables that are unrelated to others should be identified as potential outliers. To determine which objects (samples) are multivariate outliers, one should calculate a critical value by looking up critical χ^2 at the desired α-level (Tabachnick and Fidell, 1983). Confirmatory FA is performed to test the hypothesis about the structure of underlying processes (e.g. which variables in a data set form coherent clusters that are relatively independent of one another). Thus, confirmatory FA needs a validation process that can be carried out with an independent test data set (external validation) or the same data set (validation set). The algorithms used in the latter validation include cross-validation, leverage correction, bootstrap or Mallows *Cp* (Martens and Naes, 1989; Tabachnick and Fidell, 1983).

On the other hand, if the correlation matrix has variables that are 100% redundant, then the inverse of the matrix cannot be computed; it is the so-called 'ill conditioning' matrix. This happens when there are high intercorrelated variables (e.g. a variable that is the sum of two other variables). The statistical software packages can artificially add a small constant to the diagonal of the matrix, thus lowering the whole correlation in the correlation matrix, but analysts usually forget that the resulting estimates will not be exact.

In practice, there are no great differences among the methods for extracting factors. Thus, PCA could be suggested in authentication studies because it is simply a mathematical transformation of the raw data. Figure 7.2 illustrates the case of monovarietal virgin olive oils characterized by volatile compounds described in the cluster paragraph.

7.3.3 Multidimensional scaling

Multidimensional scaling (MDS) is an alternative to factor analysis when the goal of the analysis is to authenticate foodstuffs based on observed distances (similarities and dissimilarities) between investigated objects (Borg and Lingoes, 1987), in addition to correlation matrices (Shiffman *et al.*, 1981). MDS is not an exact procedure but rather a way to 'rearrange' objects in an efficient manner. The program uses a minimization algorithm that evaluates different configurations with the goal of maximizing the goodness-of-fit (or minimizing 'lack of fit'). The stress measure is used to evaluate how well (or poorly) a particular configuration reproduces the observed distance matrix. Thus, the smaller the stress value, the better is the fit of the reproduced distance matrix to the observed distance matrix. An interesting application is given in Aparicio *et al.* (1996a).

The analyst should check the Shepard diagram that represents a step line so-called 'D-hat' values. If all reproduced distances fall onto the step-line, then the rank ordering of distances (or similarities) would be perfectly reproduced by the dimensional model, while deviations from the step-line mean lack of fit. The interpretation of the dimensions usually represents the final step of this multivariate procedure. As in factor analysis, the final orientation of axes in the plane (or space) is mostly the result of a subjective decision by the researcher since the distances between objects remain invariable regardless of the type of the rotation. However, it must be remembered that MDS and FA are different methods. FA requires that the underlying data be distributed as multivariate normal, whereas MDS does not impose such a restriction. MDS often yields more interpretable solutions than FA because the latter tends to extract more factors. MDS can be applied to any kind of distances or similarities (those described in cluster analysis), whereas FA requires firstly the computation of the correlation matrix. Figure 7.3 shows the results of applying MDS to the samples described in the CA and FA sections (7.3.1 and 7.3.2).

7.3.4 Discriminant analysis

Analyses of the above type belong to the category of 'unsupervised learning', whereas discriminant analysis falls into the 'supervised' analyses of multivariate

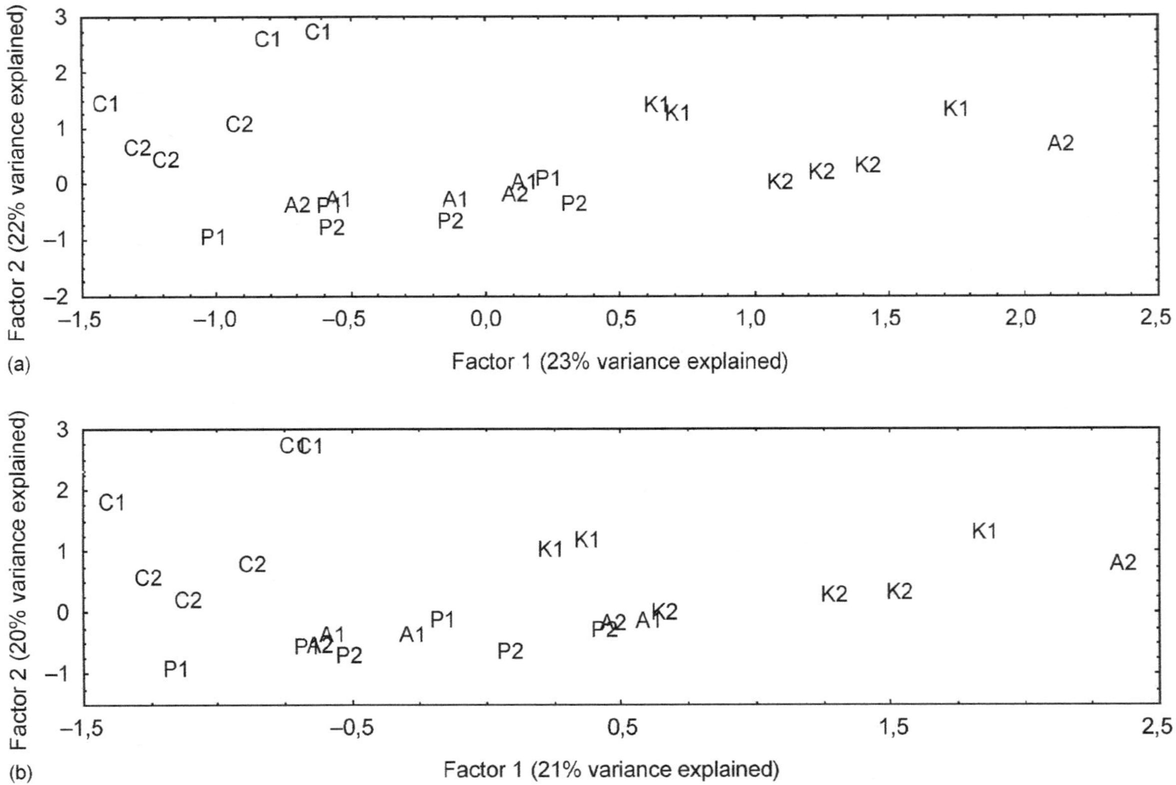

Figure 7.2 Authentication of monovarietal virgin olive oils: results of applying factor analysis to the volatile compounds. (a) Maximum likelihood and Varimax rotation. (b) Principal components and Varimax rotation. Note: A, cv. Arbequina; C, cv. Coratina; K, cv. Koroneiki; P, cv. Picual (source: SEXIA™ Group–Instituto de la Grasa, Seville, Spain).

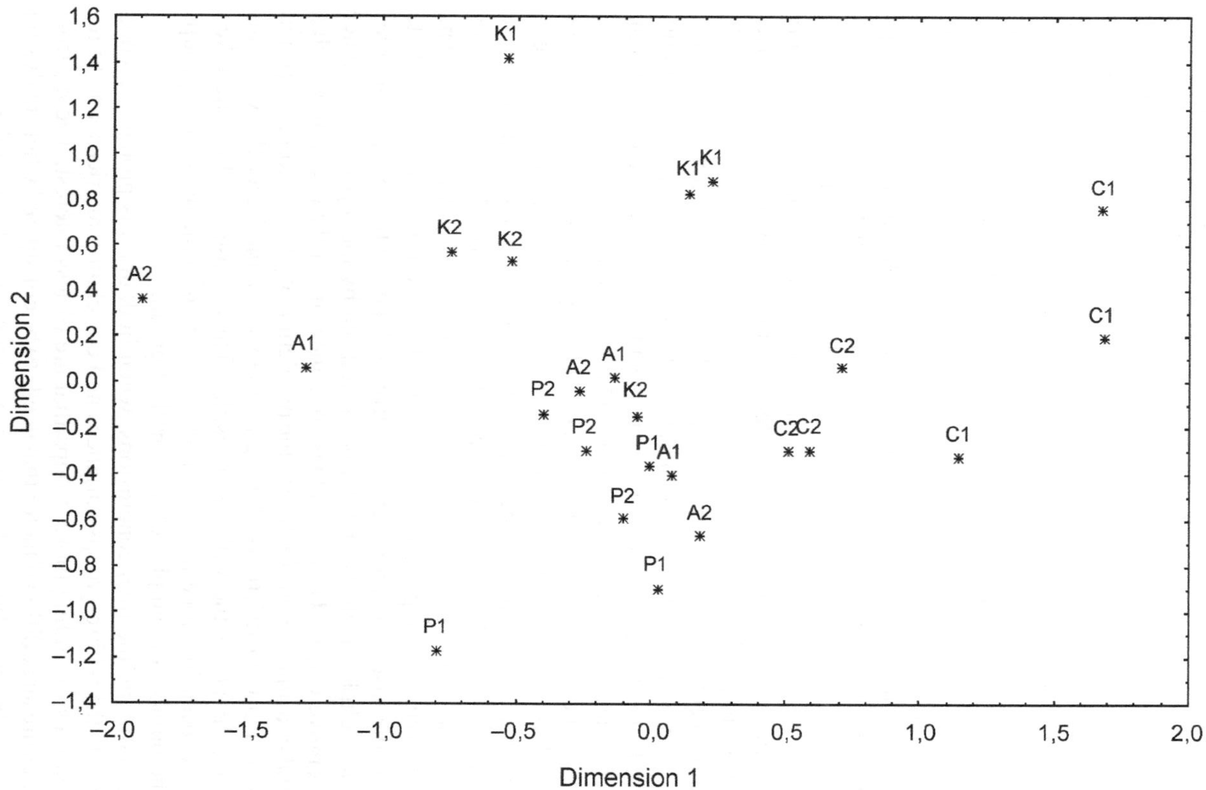

Figure 7.3 Authentication of monovarietal virgin olive oils: results of applying multidimensional scaling to volatile compounds. The distance was Manhattan (city block) and the amalgamation was Ward's method. Note: A, cv. Arbequina; C, cv. Coratina; K, cv. Koroneiki; P, cv. Picual (source: SEXIA™ Group–Instituto de la Grasa, Seville, Spain).

data. Thus discriminant analysis is used not only to determine which variables discriminate between naturally occurring groups but also which variables are the best predictors discriminating between groups. The procedure can be interpreted as a special type of factor analysis extracting orthogonal factors (Cooley and Lohnes, 1971) although it is disparaged by certain chemometricians who think that it capitalizes on chance because it 'picks and chooses' the variables to be included in the model. The procedure can render results as good as the unsupervised procedures if it is applied with rigour: Fisher for the selection of initial variables, adequate values of F-to-Enter and F-to-Remove, Jackknife algorithm (internal validation test) and an external validation set. But before applying the procedure, the analyst should inspect the means and standard deviations or variances of each group to detect outliers. They should be removed or methods used to correct their influence; a few extreme outliers have a large impact on the means and increase the variability as well. Another assumption of discriminant function analysis is that the variables used to discriminate between groups cannot be completely redundant. If any one of the variables is completely redundant with respect to the other variables then the matrix is 'ill-conditioned' and cannot be inverted.

The main objective of this procedure is to build a model in which the selected variables can predict to which group an object (sample) belongs; for example, a sample of Coratina virgin olive oil with respect to the Coratina cultivar group by a selected set of volatile compounds. The selection of variables to build the model should be done step-by-step. At each step, all unselected variables are evaluated to see which one contributes most to the discrimination between groups, and that variable is then included in the model. The backward stepwise analysis first includes all variables in the model and then, at each step, eliminates the variable that contributes least to the prediction of group membership. The model keeps only the important variables, i.e. those variables that contribute the most to the discrimination between groups. The control of the variables included or excluded from the model is carried out by the *a priori* F-to-enter and F-to-remove values. These F-values are a measure of the extent to which a variable makes a unique contribution to the prediction of group membership, and must be selected with strictness; the higher the values, the lesser the variables included but the better the validation results. Thus, those F-values are taken from an F-distribution table ($m \times n$) at 99%, where m is the number of groups and n is the number of samples of the smallest group.

Before assuming the final conclusions, when in doubt it is probably a good idea to review the within-groups variances and correlation matrices, and to re-run the analyses excluding those groups that are of less interest. The *a priori* probabilities are an additional factor that needs to be considered when analysing the objects to be authenticated; if there are more objects (samples) in one group than in any other, the discriminant analysis should be readjusted with *a priori* probabilities proportional to the sizes of the groups. Once the procedure has computed the classification scores, the analyst should know the posterior probabilities by the Mahalanobis' distance. The probability that an object can be classified into a particular group is proportional to the Mahalanobis' distance from the sample location to the group centroid.

Figure 7.4 shows the results of applying stepwise linear discriminant analysis (SLDA) to the same samples used in the previous chemometric procedures.

A good alternative to the use of supervised procedures to cluster samples, as for instance SLDA does, is to select the best variables characterising *a priori* groups by SLDA and then to run PCA exclusively with those selected variables (Baeten *et al.*, 1996). This two-step procedure would avoid building a model from pure noise or where noise had a great influence, as PCA can do under certain circumstances. Figure 7.5 shows the result of applying PCA on the four volatile compounds selected by SLDA. The groups of monovarietal virgin olive oils are much more clear after the process described than applying PCA to all volatile compounds. This means that the first two principal components of the latter procedure contained noise and hence the group classifications were worse.

7.3.5 Regression procedures

In many applications, it is expensive, time consuming or difficult to measure a property of interest directly, and the analyst should predict it based on other properties that are easier to measure. The objective is to design a model for the relationship. From the mathematical point of view, multivariate calibration is developed for finding the relationship between one or more dependent variables and a group of independent variables. In practice, a linear model is usually used for explaining the relationship but other possibilities also exist. Principal components regression (PCR), partial least squares regression (PLSR), ridge regression (RR), stepwise multiple linear regression (SMLR) and piecewise linear regression (PLR) are the most used for linear solutions. For non-linear regression logistic models, growth models, probit and logit models, among others, are used.

PCR and PLR are useful when the matrix does not contain the full model representation. The first step of PCR is the decomposition of the data matrix into latent variables through PCA and the dependent variable is then regressed onto the decomposed independent variables. PLS performs, however, a simultaneous and interdependent PCA decomposition in a way that makes that PLS sometimes handles dependent variables better than does PCR.

Ridge regression analysis is used when the independent variables are highly interrelated, and stable estimates for the regression coefficients cannot be obtained via ordinary least squares methods (Rozeboom, 1979; Pfaffenberger and Dielman, 1990). It is a biased estimator that gives estimates with small variance, better precision and accuracy.

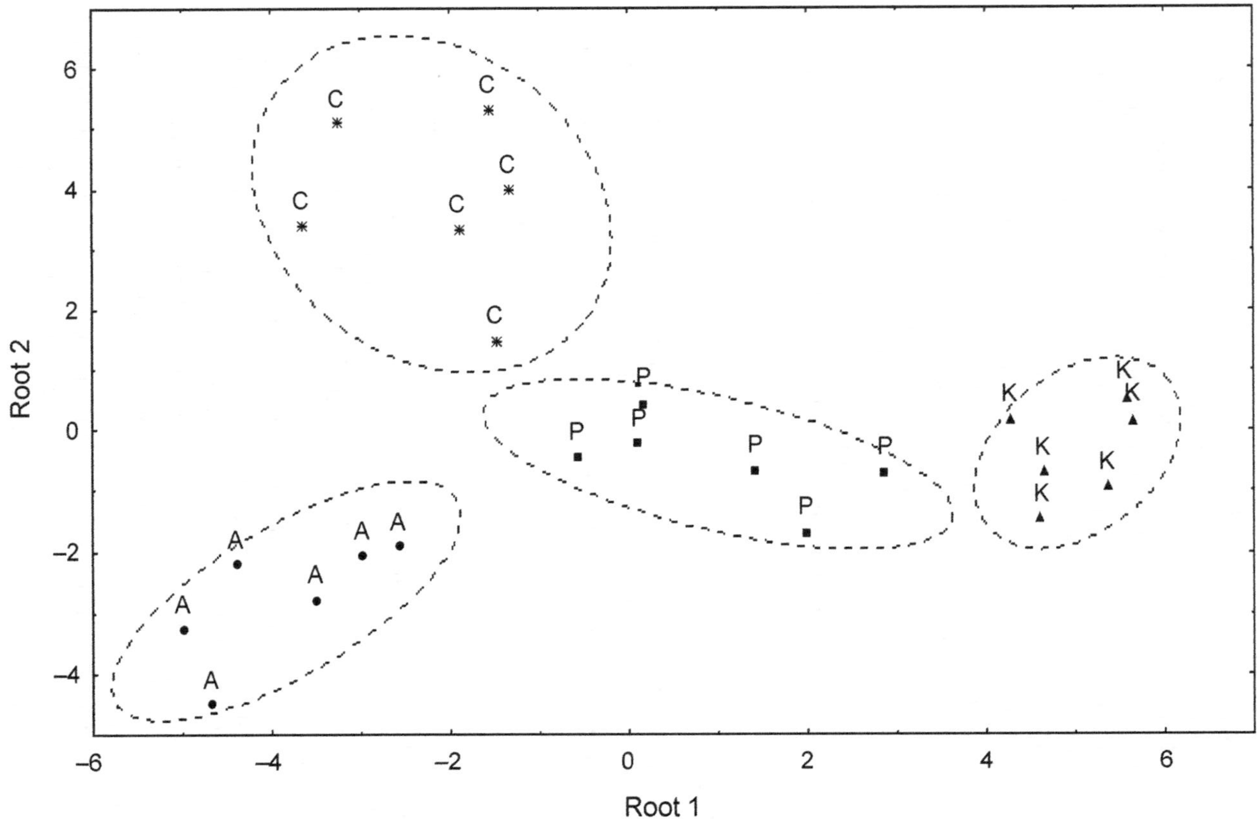

Figure 7.4 Authentication of monovarietal virgin olive oils: results of applying stepwise linear discriminant analysis to volatile compounds. Classification was carried out by four volatiles: (*E*)-2-hexenal, butyl acetate, (*E*)-3-hexenal, 2-methyl-3-buten-2-ol. *F*-to-Enter was 8.0; tolerance was upper 0.52 for all selected volatiles. Note: A, cv. Arbequina; C, cv. Coratina; K, cv. Koroneiki; P, cv. Picual (source: SEXIA™ Group–Instituto de la Grasa, Seville, Spain).

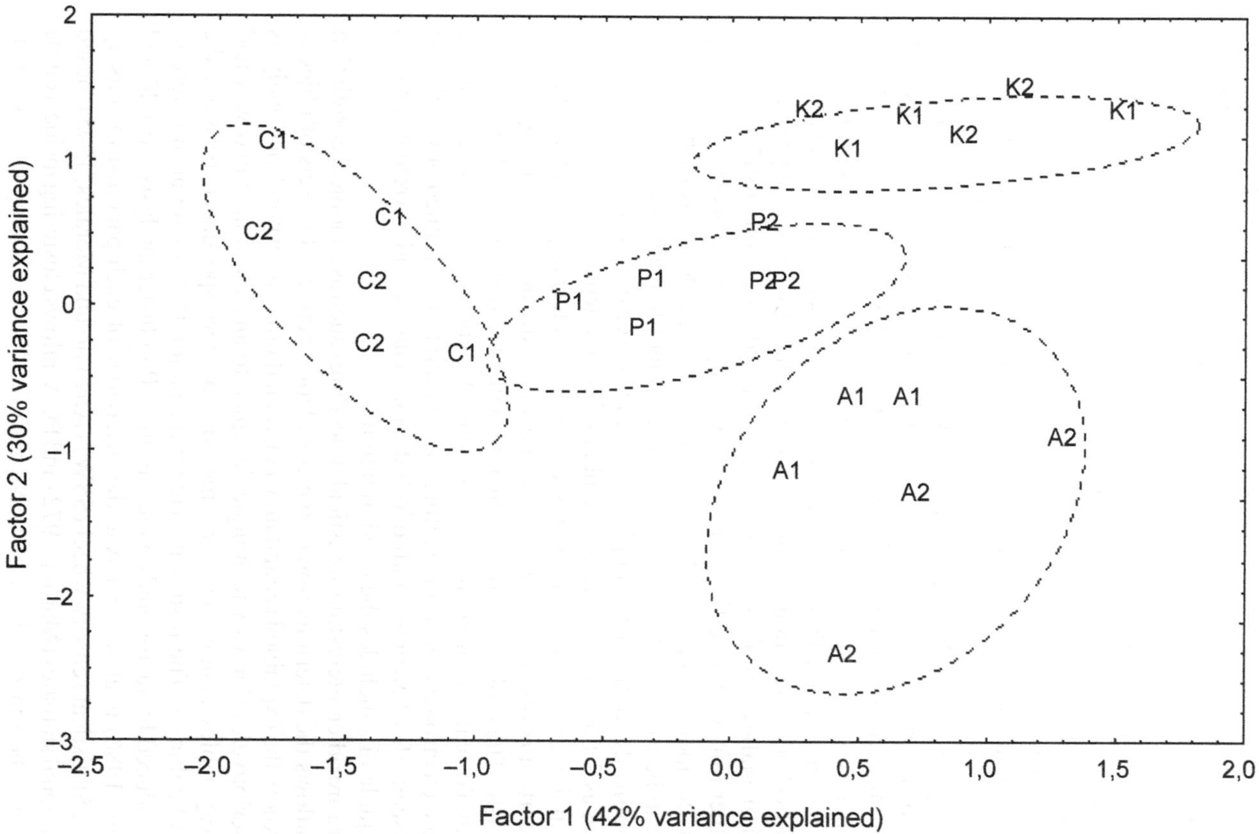

Figure 7.5 Principal components analysis applied to volatile compounds selected by stepwise linear discriminant analysis (source: SEXIA™ Group–Instituto de la Grasa, Seville, Spain).

Assumptions need to be checked concerning these regression procedures prior to their implementation. For example, it is assumed that the residuals are distributed normally. However, it is always a good idea, before drawing any final conclusions, to review the distributions of the major variables of interest, in order to inspect the distribution of the residual values (distances of the samples from the estimated model). Furthermore, residuals are also useful for detecting outliers (abnormal data) that is of special interest in authenticity studies.

Outliers are extreme samples that have distorting effects on regression analysis by causing large residual errors or having undue influence on regression coefficients. Mahalanobis' distance, evaluated as χ^2, is used to discover the outliers among samples and with respect to the solution. Outliers among variables are detected by squared multiple correlation (SMC). In this aspect, multicolinearity (this term describes two variables more or less perfectly correlated and still having similar correlations with the rest of the variables) produces high standard errors on the regression coefficient and estimation is less accurate. Furthermore, there is the singularity. This term applies when some scores are a linear combination of others. Thus, multicolinearity and singularity can cause problems in regression analyses, prohibiting or rendering matrix inversion unstable.

Another important question is how many components to use, because using too few components can build a restricted model (underfitting) whereas the model can be too flexible (overfitting) when using too many. Cross-validation or Jackknife algorithm (a re-sampling technique based on a 'leaving one sample out') can assist in determining the number of components.

The major conceptual limitation of all regression techniques is that one can only ascertain *relationships*, but one can never be sure about underlying *causal* mechanism. The explanation of conclusions with the assistance of other sciences would avoid reaching nonsense conclusions. A hypothetical paradigm can be to use the electronic nose for detecting the adulteration of refined olive oil with refined seed oils when these kinds of oils do not contain volatiles (refined process of vegetable oils includes the deodorization).

There are other interesting statistical procedures such as canonical correlation which allows the determination of the correlation between two sets of objects, and to know the explained redundancy between them. Correspondence analysis is a descriptive/exploratory technique designed to analyse simple two-way and multi-way tables containing some measure of correspondence between the rows and columns. The results provide information that is similar in nature to those produced by factor analysis techniques. Procluster analysis (Arnold and Williams, 1986) is able to analyse the behaviour of each panellist in sensory analysis. Statistical sensory wheel (SSW) (Aparicio and Morales, 1995), based on directional statistics (Mardia, 1972) and PCA, allows clustering inside a circle of not only the sensory descriptors but also the chemical compounds responsible for them. This algorithm is able to analyse the relationship between sensory descriptors and chemical compounds, certain synergies and antagonisms between chemical compounds, and the interaction between sensory descriptors and basic stimuli (Aparicio *et al.*, 1996b).

7.4 Artificial intelligence methods in food authentication

Statistical packages are generally employed for statistical production purposes because they provide the users with the mechanics of the data analysis but they do not help very much with analytical strategies. Thus, when the analyst faces the authentication problem and sees that all the measures have an uncertainty associated with measurement results, s/he can make the decision of working with methods of artificial intelligence in which the inexactness (generality, ambiguity, vagueness and uncertainty) is the main reason for their existence. The normal question is to wonder which method (neural network, fuzzy logic and expert systems) will give the best conclusions. Each one of these methods (models) has its advantages and disadvantages. Although they are dissected below, simple definitions could enlighten us about their possible application in food authentication. Neural networks are models that can learn from past experiences by adaptive programs, which means that they have some of the disadvantages of the discrimination systems. The theory of fuzzy sets provides a convenient means to deal with ill-defined and doubtful data. Expert systems are self-learning, efficient and reliable programs that are able to solve a particular problem based on knowledge bases of heuristic rules.

7.4.1 Expert systems

The ability to make decisions on the basis of knowledge distinguishes expert systems (ESs) from statistical programs but their essential characteristics are the self-learning, the look-ahead and the back-propagation up to the point that they are *sine qua non* conditions for true expert systems. Neither statistical procedures nor other artificial intelligence algorithms are able to learn from past experiences, evaluate decisions before making them (chess's rules) or retry if a decision led the system in a wrong way. Expert systems have to be designed to carry out two main tasks: (i) to extract knowledge rules, and (ii) to design the algorithm controlling them. An ES comprises a knowledge base and an inference engine (Klahr and Waterman, 1986). The former stores the domain knowledge in the form of facts and rules, whereas the latter is able to cluster only the knowledge related to a specific knowledge domain (e.g. a monovarietal virgin olive oil, the geographical origin of a virgin olive oil) and to apply the rules in the appropriate order to infer the correct conclusion. The better the selected information, the greater the experts system success.

Expert systems are based on empirical data and, obviously, the data set should fulfil certain conditions if the process is to be successful; an expert system based on non-refined data would be meaningless. The most important of these conditions is related to the distribution of values over the years and, frequently, to the degree of separability of the foods to be authenticated. The empirical information is used to build some of the knowledge rules.

Expert systems work with rules that have been built independently. Each rule has propositions that are related to a parameter, or a ratio between parameters or an equation classifying categories (e.g. two monovarietal virgin olive oils) of the knowledge domain (e.g. cultivars) to be characterized. The taxonomic organization of the knowledge takes an arborescent form or tree graph (Aparicio, 1988; Zupan *et al.*, 1988). Each node contains information about the class finding and this information is held in structures called frames. A frame is an abstract specification for a class similar to a property list or record in conventional programming. The frames contain two types of information: domain-specific components and external information. The domain-specific component expresses the substantive characteristics of each parameter while the latter is associated with facts given by the user. Each entity of a frame consists of a name, its attributes and the values linked with these attributes (e.g. *superclass*: aliphatic alcohols; *class*: tetracosanol; *concentration range*: 1.33, 20.37; *geographical origin*: Andalusia). The ES knowledge base is represented by rules obtained from the frame. A rule is represented by means of a set of propositions (premises) that, if they are fulfilled, provides a conclusion associated with a value that indicates its certainty factor (CF): RULE ⟨rule name⟩, IF ⟨propositions⟩, THEN ⟨conclusion⟩; CF= ⟨value⟩.

There are different types of knowledge rules, though the most common are four (Aparicio, 1988): inexact reasoning rules, lineal rules, relational rules and heuristic rules. The first type of rules stores the distinctive values of chemical compounds that are characteristics of each cluster of the knowledge domain (e.g. cv. Picual) as premises (e.g. IF the tridecene concentration is high, THEN the cultivar is Hojiblanca; CF = 0.78) The second type of rules stores the coefficients of lineal equations obtained by statistical procedures working with bivariate distributions. The boundaries of each one of these distributions draw a geometrical figure, so-called 'fuzziness-triangle', where the peak point corresponds to the maximum probability calculated by statistical procedures (e.g. tridecene concentration (ppm) high [0.612, 0.669, 0.963]; medium [0.342, 0.510, 0.669]; low [0.178, 0.342, 0.510]). The relational rules are quite similar to the previous one although they do not implement an arithmetic calculation among chemical parameters (e.g. IF %Oleic is less than %Linolenic $*$ 100, THEN the geographical origin is Andalusia; CF = 0.81) (Alonso and Aparicio, 1993). The last rule assists the system to determine whether the conclusions pointed out from different frames (e.g. cultivar, geographical origin, extraction systems) are in agreement with one another; for example, if the conclusions are 'cv. Koroneiki' (cultivar) and 'Greece' (geographical origin), the results would be consistent enough because Koroneiki is a Greek cultivar. If the variety were 'cv. Picual' (Spanish cultivar) there would be no agreement and the system would try to find

Table 7.1 Authentication of the geographical origin of virgin olive oil samples: comparative results of SEXIA™ expert system, neural networks and the supervised chemometric procedure of stepwise linear discriminant analysis. Samples collected in the regions of Jaén (Spain)

		Correct classifications (%)		
Region	Number of samples	SEXIA	SLDA	ANN
Campiña	16	99.9	62.5	94.1
Cazorla	8	87.5	25.0	75.0
Condado	8	99.9	75.0	87.5
La loma	17	94.1	82.4	94.1
Martos	19	94.7	84.2	89.5
Sierra Morena	6	99.9	83.3	83.3
Sierra del Segura	12	83.3	58.3	91.7
Sierra Sur	11	90.9	18.2	91.9

other conclusions by following alternative routes in the tree structure (Bundy, 1983; Shirai and Tsujii, 1982).

SEXIA™ is the expert system designed to authenticate virgin olive oil in terms of its geographical origin and cultivars (Aparicio, 1988; Aparicio and Alonso, 1994; Alonso and Aparicio, 1993; Aparicio *et al.*, 1994). To date, its knowledge base contains more than 400 reasoning rules that compile around 2000 European virgin olive oil samples characterized by 55 chemical compounds. A comparison between the results attained by statistical procedures and this expert system showed that SEXIA™ was 26% better than discriminant analysis in the studies of virgin olive oils produced in Italy (Aparicio *et al.*, 1994). Table 7.1 shows the results of authenticating the geographical origin of virgin olive oil samples produced by the cooperative societies located in the eight regions of the province of Jaén (Spain) that is the highest world producer with an average of about 400,000 tons (1996–1997). Other expert systems have been used or designed by Blaffert (1986), Derde *et al.* (1987), Betteridge *et al.* (1988) and Adler *et al.* (1993).

7.4.2 Neural networks

Neural networks are essentially non-linear regression models based on a binary threshold unit (McCulloch and Pitts, 1943). The structure of neural networks, called a perceptron, consists of a set of nodes at different layers where the node of a layer is linked with all the nodes of the next layer (Rosenblatt, 1962). The role of the input layer is to feed input patterns to intermediate layers (also called hidden layers) of units that are followed by an output result layer where the result of computation is read-off. Each one of these units is a 'neuron' that computes a weighted sum of its inputs from other 'neurons' at a previous layer, and outputs a one or a zero according to whether the sum is above or below a certain threshold. The weighted sum of inputs must be or exceed the threshold for a unit to fire. Thus, zero means no connection between units whereas positive or negative values mean an excitatory or inhibitory connection between units. This classic definition is generalized in the following equation

$$n_i := g(\Sigma w_{ij} n_j - \mu_i), \text{ for } j = 1 \text{ to } n$$

where n_i represents the state of the unit, $g(x)$ is a general non-linear function, w_{ij} represents the strength of the connection between unit i and unit j, μ_i is the specific threshold value for unit i, and the sign := emphasises that the equation is not a function of time since it is updated asynchronously by a common clock.

The question is: how does the analyst choose w_{ij} so that a model can carry out a task? The solution can be the iterative adjustments of the w_{ij} strengths that may be done by supervised or unsupervised learning. The iterative learning procedure based on supervised learning is stopped either after a predetermined number of cycles or after a defined minimum difference between the network output and the correct outputs. For those situations in which the information available is extremely reduced, there are particular modelling networks for training stage of the neural networks, for example 'learning with a critic' (Barto *et al.*, 1991) and 'reward–penalty' (Barto and Anandan, 1985).

The unsupervised procedure does not have a feedback saying how the outputs should be or whether they are correct. This means that unsupervised learning can only do anything useful when there is redundancy in the input data (Barlow, 1989). The unsupervised learning can be based on multiple output units that are often active together (e.g. Hebbian learning and extension rules) (Hebb, 1949; Oja, 1989; Sanger, 1989) or on only one output unit per group at a time (e.g. competitive learning). The former learning is very similar to the statistical algorithms of principal components and cluster analysis whereas competitive learning categorizes the input data by models such as Willshaw and von der Masburg or Kohonen (Hertz *et al.*, 1991; Kohonen, 1989). The unsupervised learning is rapid as it does not use back-propagation (which can be extremely slow) and it is advisable to apply it before training a network with a back-propagation supervised procedure.

Another division of neural networks corresponds to the number of layers: a simple perceptron has only one layer (Minski and Papert, 1969), whereas a multilayer perceptron that has more than one layer (Hertz *et al.*, 1991). This simple differentiation means that network architecture is very important and each application requires its own design. To get good results one should store in the network as much knowledge as possible and use criteria for optimal network architecture as the number of units, the number of connections, the learning time, cost and so on. A genetic algorithm can be used to search the possible architectures (Whitley and Hanson, 1989).

Table 7.1 compares the results of authenticating the geographical origin of virgin olive oil samples produced in the regions of Jaén (Spain) by an expert sytem, the supervised procedure of SLDA and the neural network.

7.4.3 *Fuzzy logic*

In many experimental cases, a certain degree of interference occurs among the measures, which gives rise to possible collections of results; however, the situation is even more complex if the input data are subjected to uncertainty or imprecision (Kaufmann and Gupta, 1991). Fuzzy logic is the only mathematical application that can properly solve problems with imprecision in input data.

Fuzzy logic is based on the generalization of theory of sets 'characteristic function' that Zadeh defined as 'membership function', $\mu(x)$, (Zadeh, 1965)

$$\mu_F(x_i) \in [0, 1] \quad \forall x_i \in F$$

The function describes the degree of membership, within the closed interval [0,1], to which each element x_i in the knowledge domain belongs to a fuzzy subset, F. The degree of membership 1 means that that x_i definitely belongs to F (full membership), zero means that it does not belong to the fuzzy subset (non-membership), whereas a value between zero and one means partial membership. The relationship between possible values of x_i and their degree of membership can be represented by different graphs as triangular, trapezoidal, bell-shaped or irregular asymmetric functions (Kandel, 1986). Based on this function, one can build models of input data into degrees of membership of linguistic fuzzy sets, i.e. low, medium or high; this is called 'fuzzification'. We can further define special modifiers that would act to modify the initial probability distributions; for example:

$$\text{Very X: } \mu_{\text{very}}\text{X}(\text{F}_i) = \mu^2(\text{F}_i)$$

and

$$\text{Fairly X: } \mu_{\text{more or less}}(\text{F}_i) = \mu^{1/2}(\text{F}_i)$$

But in order to build a practical system that must be precise and consistent in its behaviour, one has to impose restrictions with regard to the use of ambiguous modifiers (e.g. very not large) and the order of them in the composition (e.g. very X, fairly X, more or less X, above X, much above X) (Dubois and Prade, 1980; Kandel, 1986).

However, the identification of the fuzziness associated with single parameter characterizing foodstuffs is not enough. The authentication process is not usually restricted to a single parameter but in fact there are often several of them. If we want to operate with their membership function (e.g. low linoleic and high 24-methylene cycloarthanol), we need to define operations on the fuzzy set. Thus, the classical rule 'R = IF (input) A, THEN (output) B' can be extended

Table 7.2 Authentication of monovarietal virgin olive oils: comparative results of fuzzy logic algorithms (Calvente and Aparicio, 1995) and the supervised chemometric procedure of linear discriminant analysis. Chemical compounds used: linolenic acid, 24-methylencycloarthanol sterol and copaene hydrocarbon

		Correct classifications (%)	
Cultivar	Number of samples	Fuzzy logic	LDA
Farga	24	91.7	91.7
Hojiblanca	35	62.9	74.3
Picual	228	93.9	96.9

to several inputs (antecedents), and perhaps outputs (consequences), linked by the aforementioned logic operations; for example, IF (fairly low linoleic fatty acid) and (very high 24-methylene cycloarthanol), THEN variety is Arbequina (Calvente and Aparicio, 1995). The fuzzy relation between pair of sets can also be expressed using basic operations as T-norm, T-conorm, negation, implication and defuzzification (Kandel, 1986)

Table 7.2 shows the results of comparing the results of authenticating three Spanish monovarietal virgin olive oils by the supervised procedure of linear discriminant analysis and an algorithm of fuzzy logic. The results are promising as the results applying Luckasiewicz T-conorm $S_{1.5}$ (Kandel, 1986) are of the same order of magnitude than linear discriminant analysis.

References

Adler, B., Schütze, P. and Will, J. (1993) Expert system for interpretation of x-ray diffraction spectra. *Anal. Chim. Acta*, **271**, 287–291.

Alonso, V. and Aparicio, R. (1993) Characterization of European virgin olive oils using fatty acids. *Grasas Aceites*, **44**, 18–24.

Aparicio, R. (1988) Characterization of foods by inexact rules: The SEXIA expert system. *J. Chemometr. A*, **3**, 175–192.

Aparicio, R. and Alonso, V. (1994) Characterization of virgin olive oils by SEXIA expert system. *Prog. Lipid Res.*, **33**, 29–38.

Aparicio, R. and Morales, M.T. (1995) Sensory wheels: a statistical technique for comparing QDA panels. Application to virgin olive oil. *J. Sci. Food Agric.*, **67**, 247–257.

Aparicio, R., Alonso, V. and Morales, M.T. (1994) Detailed and exhaustive study of the authentication of European virgin olive oils by SEXIA expert system. *Grasas Aceites*, **45**, 241–252.

Aparicio, R., Calvente, J.J. and Morales, M.T. (1996a) Sensory authentication of European extra-virgin olive oil varieties by mathematical procedures. *J. Sci. Food Agric.*, **72**, 435–447.

Aparicio, R., Morales, M.T. and Alonso, V. (1996b) Relationship between volatile compounds and sensory attributes by statistical sensory wheel. *J. Am. Oil Chem. Soc.*, **73**, 1253–1264.

Aparicio, R., Morales, M.T. and Alonso, V. (1997) Authentication of Europen virgin olive oils by their chemical compounds' sensory attributes and consumers' attitudes. *J. Agric. Food Chem.*, **45**, 1076–1083.

Aparicio, Morales, M.T., Luna, G. and Aparicio-Ruiz, R. (2000) Biochemistry and chemistry of volatile compounds affecting to consumers' attitudes of virgin olive oil, in *Flavour and Fragrance Chemistry* (eds V. Lanzotti and O. Taglialatela-Scafati), Kluwer Academic Press, Dordrecht, The Netherlands, pp. 3–14.

Armstrong, R.D. and Beck, P.O. (1990) An algorithm to assist in the identification of multiple multivariate outliers when using a least absolute value criterion, in *Robust Regression: Analysis and Applications* (eds K.D. Lawrence and J.L. Arthur), Marcel Dekker, New York, pp. 89–104.

Arnold, G.M. and Williams, A.A. (1986) The use of generalised procrusters and techniques in sensory analysis, in *Statistical Procedures in Food Research* (ed. J.R. Piggott), Elsevier, London, pp. 233–254.

Baeten, V., Meurens, M., Morales, M.T. and Aparichio, R. (1996) Detection of virgin olive oil adulteration by fourier transform raman spectroscopy. *J. Agric. Food Chem.*, **44**, 2225–2230.

Barlow, H.B. (1989) Unsupervised learning. *Neural Comput.*, **1**, 295–311.

Barto, A.G. and Anandan, P. (1985) Pattern recognizing stochastic learning automata. *IEEE Trans. Syst. Man Cyber.*, **15**, 360–375.

Barto, A.G., Sutton, R.S. and Watkins, C.J.C.H. (1991) Learning and sequential decision making, in *Learning and Computational Neuroscience* (eds M. Gabriel and J.W. Moore), MIT Press, Cambridge, MA, USA, pp. 112–136.

Betteridge, D., Mackison, R., Mottershead, C.M., Taylor, A.F. and Wade, A.P. (1988) Development of an expert system for the selection of sample points for moisture analysis. *Anal. Chem.*, **60**, 1534–1539.

Blaffert, T. (1986) EXPERTISE—an expert system for infrared spectra evaluation. *Anal. Chim. Acta*, **191**, 161–168.

Borg, I. and Lingoes, J. (1987) *Multidimensional Similarity Structure Analysis*, Springer Publishing Co., New York.

Bundy, A. (1983) *The Computer Modelling of Mathematical Reasoning*, Academic Press, London.

Calvente, J.J. and Aparicio, R. (1995) A fuzzy filter for removing interferences among membership grade functions. An application to pre-treatment of data in olive oil authentication. *Anal. Chim. Acta.*, **312**, 281–294.

Cooley, W.W. and Lohnes, P.R. (1971) *Multivariate Data Analysis*, John Wiley and Sons, New York.

Derde, M.P., Buydens, L., Guns, C. and Massart, D.L. (1987) Comparison of rule-building expert systems with pattern recognition for the classification of analytical data. *Anal. Chem.*, **59**, 1868–1871.

Dubois, D. and Prade, H. (1980) *Fuzzy Sets and Systems: Theory and Applications*, Academic Press, San Diego, CA, USA.

EURACHEM (1998) The fitness for Purpose of Analytical Methods. A Laboratory Guide to method Validation and Related Topics. EURACHEM Secretariat, Teddington, UK.

Hartigan, J.A. (1975) *Clustering Algorithms*, John Wiley and Sons, New York.

Hebb, D.O. (1949). *The Organization of Behaviour*, John Wiley and Sons, New York.

Hertz, J., Krogh, A. and Palmer, R.G. (1991) *Introduction to the Theory of Neural Computation*, Addison-Wesley Publishing Co., Reading, MA, USA.

Jacobsen, T. and Gunderson, R.W. (1986) Applied cluster analysis, in *Statistical Procedures in Food Research* (ed. J.R. Piggot), Elsevier Applied Science, London, pp. 361–408.

Kandel, A. (1986) *Fuzzy Mathematical Techniques with Applications*, Addison-Wesley Publishing Co., Reading, MA, USA.

Kaufmann, A. and Gupta, M.M. (1991) *Introduction to Fuzzy Arithmetic: Theory and Applications*, van Nostrand Reinhold, New York.

Klahr, P. and Waterman, D.A. (1986) *Expert System Techniques, Tools and Applications*, Addison-Wesley Publishing Co., Reading, MA, USA.

Kohonen, T. (1989) *Self-Organization and Associative Memory*, Springer-Verlag, Berlin, Germany.

Mardia, K.V. (1972) *Statistics of Directional Data*, Academic Press, New York.

Martens, H. and Naes, T. (1989) *Multivariate Calibration*, John Wiley and Sons, Chichester, UK.

Massart, D.L. and Kaufman, L. (1983) Hierarchical clustering methods. In *The Interpretation of Analytical Chemical Data by the Use of Cluster Analysis* (ed. J.D. Winefordner), John Wiley and Sons, New York, pp. 75–99.

Minski, M.L. and Papert, S.A. (1969) *Perceptions*, MIT Press, Cambridge, MA, USA.

Oja, E. (1989) A simplified neuron model as a principal component analyzer. *J. Math. Biol.*, **15**, 267–273.

Pfaffenberger, R.C. and Dielman, T.E. (1990) A comparison of regression estimators when both multicollinearity and outliers are present, in *Robust Regression: Analysis and Applications* (eds K.D. Lawrence and J.L. Arthur), Marcel Dekker, New York, pp. 243–270.

Piggot, J.R. and Sharman, K. (1986) Method to aid interpretation of multidimensional data, in *Statistical Procedures in Food Research* (ed. J.R. Piggott), Elsevier Applied Science, London, pp. 181-132.

Rosenblatt, F. (1962) *Principles of Neurodynamics*, Spartan, New York.

Rozeboom, W.W. (1979) Ridge regression: bonanza or beguilment? *Psychol. Bull.*, **86**, 242–249.

Sanger, T.D. (1989) Optimal unsupervised learning in a single-layer linear feedforward neural network. *Neural Networks*, **2**, 359–473.

Schiffman, S., Reynolds, M.L. and Young, F.W. (1981) *Introduction to Multidimensional Scaling. Theory, Methods and Applications*, Academic Press, Orlando, FL, USA.

Shirai, Y. and Tsujii, J. (1982) *Artificial Intelligence: Concepts, Techniques and Applications*, John Wiley and Sons, Chichester, UK.

Tabachnick, B.G. and Fidell, L.S. (1983) *Using Multivariate Statistics*, Harper and Row, New York.

Taylor, B.N. and Kuyaat, C.E. (1994) Guidelines for evaluating and expressing the uncertainty of NIST measurement results, National Institute of Standards and Technology, Gaithersburg, MD, USA.

Wherry, R.J. (1984) *Contributions to Correlational Analysis*, Academic Press, New York.

Whitley, D. and Hanson, T. (1989) Optimizing neural networks using faster more acurate genetic search. *Proceedings of Third International Conference on Genetic Algorithms* (eds J.D. Schaffer, C.A. San Mateo, and M. Kaufmann), pp. 391–396.

Wold, S. (1978) Cross-validatory estimation of the number of components in factor analysis and principal components models. *Technometrics*, **20**, 397–406.

Zadeh, L.A. (1965) Fuzzy sets. *Inf. Control*, **8**, 338–353.

Zupan, J. (1982) *Clustering of Large Data Sets*, Research Studies Press, New York.

Zupan, J., Razinger, M., Bohanec, S., Novic, M., Tasar, M. and Lah, L. (1988) Building knowledge into an expert system. *Chem. Int. Lab. Syst.*, **4**, 307–314.

Zupan, J., Novic, M., Li, X. and Gasteiger, J. (1994) Classification of multicomponent analytical data of olive oils using different neural networks. *Anal. Chim. Acta*, **292**, 219–234.

8 Authenticity of edible oils and fats: the legal position*

Catriona Stewart

8.1 Introduction

Food authenticity may, in simple terms, be taken to mean that the product purchased by the consumer matches its description. If it does not, and the food is misdescribed or mislabelled, the consumer may be misled. Misdescription or mislabelling, as they relate to edible fats and oils, encompass a range of issues. The main possibilities, however, are the dilution/adulteration/substitution of one oil with another, and false claims regarding geographical or regional origin. The basis of the problem in both cases is, essentially, that some oils trade at a premium in comparison with others. Peanut oil (arachis oil), for example, tends to be more expensive than rapeseed oil or soyabean oil. Olive oil (particularly the higher quality grades) and some of the speciality oils are more expensive again. Within the extra virgin and virgin olive oil sectors, an additional market force comes into play – the country or geographical region of origin. Consumers are prepared to pay more for an oil from a particular region than for a product which is a blend of oils from a number of countries. The potential profits and trading advantages to be made from mislabelling the oil type or place of origin may, thus, be very considerable. Although these issues, in the main, are unlikely to have safety implications, such practices prejudice the interests of both consumers and honest traders, and are contrary to food law.

In the UK, enforcement of most of the relevant food law is the responsibility of local authorities and port health authorities. The Food Standards Agency too has a regulatory responsibility to protect consumers from food mislabelling. To assist with this function the Agency has an ongoing research programme to develop new tests for assessing the authenticity of foods and a surveillance programme to check, through small-scale surveys, that foods are being correctly described and labelled.

This chapter outlines the legislation that has been developed at the UK national level, at European level and internationally with respect to the authenticity of fats and oils. It then goes on to describe briefly the authenticity research

---

*The views expressed in this chapter are those of the author and should not be regarded as a statement of official UK Government policy.

and development, and surveillance programmes of the Food Standards Agency, focusing on the work related to vegetable oils.

8.2 UK and European legislation

With the exception of olive oils, there is no specific legislation in the UK controlling the labelling of edible fats and oils. A measure of general protection from misdescription is, however, given by the *Trades Description Act 1968* (UK Parliament, 1968). More specific protection is afforded by the *Food Safety Act 1990* (UK Parliament, 1990) and the *Food Labelling Regulations 1996* (GB Statutory Instrument, 1996), as amended. This legislation applies in Great Britain while separate, but similar, legislation applies in Northern Ireland.

8.2.1 Trades Description Act 1968

This Act applies to all articles on sale and makes it an offence to apply a false 'trades description' to any goods. The term 'trades description' covers among other things: the quality of the goods; how the goods are made or processed; what the goods are made of; and, where the goods were made. The information must be false to a material degree for there to be an offence. It must be applied to the goods in question, whether in writing or by means of an illustration, symbol or other marking on the goods themselves, on containers, labels, show cards, in advertisements, etc. or in an oral statement.

8.2.2 Food Safety Act 1990 and Food Labelling Regulations 1996

Under Sections 14 and 15 of the *Food Safety Act*, it is an offence to sell food that is not of the nature, substance or quality demanded by the consumer or to falsely or misleadingly describe or present food. The presence of any undeclared oils or false claims regarding the geographical origin of the oil may be considered to constitute such an offence.

The Food Safety Act also provides the enabling powers under which most other UK food regulations, including the *Food Labelling Regulations 1996* (as amended), are made. These Regulations implement Council Directive 79/112/EEC that has now been superseded by European Parliament and Council Directive 2000/13/EC (European Communities, 1978, 2000) on food labelling. The principal measures forming the pillars of consumer protection with respect to authenticity are the requirements that most products must be labelled with the 'name of the food' and must include a list of ingredients.

8.2.2.1 Name of the food

With regard to the 'name of the food', where there is a name laid down by law (names which either European or UK law specifies must be used for certain foods), this must be used. If not, a customary name (a name that, in time, has come to be accepted by consumers as the name of a food without it needing further explanation) may be used. If there is no customary name, or it is not used, a descriptive name must be used. A descriptive name must be precise enough to indicate the nature of the product and to distinguish it from other foods with which it could be confused. Descriptive names are generally used for vegetable oils, e.g. 'corn oil', the exception being for olive oils (the relevant legislation on this will be dealt with in the next section). An oil labelled as 'corn oil', but to which rapeseed oil has been added but not declared, is misleading to the consumer and in contravention of the rules.

8.2.2.2 Ingredients listing

The provisions relating to the name of the food are clearly fundamental to consumer protection with respect to mislabelled cooking fats and oils which are single ingredient foods. For foods containing fats and oils, another important aspect of control is the requirement that the label of most foods must include a list of ingredients. Again, though, there is potential for financial gain or trading advantage by mislabelling of this information.

The name used for an ingredient should be the name, including legal names, which could be used for it if it were being sold as a food by itself and it should reflect its true nature. If the term 'olive oil', for example, is used in the ingredients listing of a spreadable fat then that oil must comply with the description and criteria specified for 'olive oil' in the relevant marketing standards (see section 8.2.3). Use of cheaper 'olive-pomace oil' instead would be contrary to the requirements. Certain generic terms are permitted for ingredients and this includes 'oil', which may be used for any refined oil other than olive oil, and 'fat', which may be used for any refined fat. Notwithstanding this, in order to give the consumer information on the source, the generic name must be accompanied by either the description 'animal' or 'vegetable', as appropriate, or an indication of the specific animal or vegetable species.

The name of an ingredient should include appropriate references to the physical condition or to any process or treatment that it has undergone in cases where omission of this information would mislead. Therefore, if oil or fat used as an ingredient in any product has been hydrogenated, a declaration of this must be included. Hydrogenation alters the composition of the oil and results in a decrease in the content of polyunsaturated fatty acids and an increase in monounsaturated, saturated and *trans* fatty acids (TFAs). Consumers wishing to avoid products with a potentially high TFA content may be misled if this aspect of the processing of the oil or fat component has not been highlighted.

The Regulations also require a quantitative ingredient declaration (QUID) generally to be given for those ingredients that:

- appear in the name of a food or which are usually associated with that name by consumers
- are emphasized on the label in either words, pictures or graphics
- are essential to characterize the food and to distinguish it from products with which it might be confused because of its name or appearance.

The declaration must be shown as a percentage, must appear in or next to the name of the food (or in the ingredients list), and will usually be based on the amount of the ingredient used in the preparation of the food. A spreadable fat marketed on the basis that it contains olive oil, for example, should provide a quantitative indication of how much olive oil has been used in its manufacture. Omission of this information could be misleading as the consumer may assume justifiably that the entire fat of the product is derived from olive oil when this is, in fact, not the case.

8.2.2.3 Nutrition labelling

Nutrition labelling, where it is given, provides a valuable contribution to consumer information and consumer choice. It is voluntary to provide this information unless a nutrition claim (e.g. high in polyunsaturates) is made. Nutrition labelling must include the amount of any nutrient for which a claim is made. As a minimum, the energy, protein, carbohydrate and fat must be listed (this is known as a Group 1 declaration). However, the UK Government recommends that energy, protein, carbohydrate, sugar, fat, saturates, fibre and sodium are given (Group 2 declaration) as this gives consumers information on the key health-related nutrients (MAFF, 1999). If a claim is made regarding the content of monounsaturates, polyunsaturates or cholesterol, the saturates must also be declared.

8.2.2.4 Other labelling requirements

In addition to the name of the food and the ingredients listing, information must also be provided on the food's weight or volume, how long it will keep for, the name and address of the manufacturer, packer or seller and, in some cases, where the product was made, and how to store and prepare it. The aim of this information is to help consumers to compare products that may seem similar and help them to decide which one to buy. Giving false information in respect of these requirements is, therefore, potentially misleading.

8.2.3 Marketing standards for olive oil

The situation for olive oils is different to that for other vegetable oils. Council Regulation 136/66/EEC, as amended by Council Regulations 356/92 and 1638/98 (European Communities, 1966, 1992a, 1998a), names and defines nine

Table 8.1 Descriptions and definitions of olive oils and olive-pomace oils

Description	Definition
Virgin olive oils	This is a generic term and refers to oils from the fruit of the olive tree derived solely by mechanical or other physical means under conditions, particularly thermal conditions, that do not lead to alteration of the oil, and which have undergone no treatment other than washing, decantation, centrifugation or filtration, but excluding oils obtained using solvents or re-esterification processes and any mixtures with oils of other kinds.
Extra virgin olive oil*	Virgin olive oil having a maximum free acidity, expressed as oleic acid, of 1 g per 100 g and the other characteristics laid down for this category.
Virgin olive oil*	Virgin olive oil having a maximum free acidity, expressed as oleic acid, of 2 g per 100 g and the other characteristics laid down for this category.
Ordinary virgin olive oil	Virgin olive oil having a maximum free acidity, expressed as oleic acid, of 3.3 g per 100 g and the other characteristics laid down for this category.
Lampante virgin olive oil	Virgin olive oil having a free acidity, expressed as oleic acid, of greater than 3.3 g per 100 g and/or the other characteristics laid down for this category.
Refined olive oil	Olive oil obtained by refining virgin olive oil, having a maximum free acidity, expressed as oleic acid, of 0.5 g per 100 g and the other characteristics laid down for this category.
Olive oil*	Olive oil obtained by blending refined olive oil and virgin olive oil, other than lampante oil, having a maximum free acidity, expressed as oleic acid, of 1.5 g per 100 g and the other characteristics laid down for this category.
Crude olive-pomace oil	Oil obtained by treating olive pomace with solvents, excluding oil obtained by means of re-esterification and mixtures with other types of oil, and having the other characteristics laid down for this category.
Refined olive-pomace oil	Oil obtained by refining crude olive-pomace oil, having a maximum free acidity, expressed as oleic acid, of 0.5 g per 100 g and the other characteristics laid down for this category.
Olive-pomace oil*	Oil obtained by blending refined olive-pomace oil and virgin olive oil, other than lampante oil, having a maximum free acidity, expressed as oleic acid, of 1.5 g per 100 g and the other characteristics laid down for this category.

*Only these categories of olive oil may be sold at retail level.

categories of olive oil and olive-pomace oil (formally referred to as 'olive-residue oil'). These are listed in Table 8.1 and only these categories may be marketed within the European Community. Only four of these may be sold at retail level, i.e. 'extra virgin olive oil', 'virgin olive oil', 'olive oil' and 'olive-pomace oil'. Refined olive oil and refined olive-pomace oil must be blended with virgin oils before retail sale. These council regulations are directly applicable in all Member States. In Great Britain, enforcement provisions are contained in the *Olive Oil (Marketing Standards) Regulations 1987* (GB Statutory Instrument, 1987), as amended. Separate, but similar, legislation applies in Northern Ireland.

Table 8.2 Physicochemical criteria for distinguishing olive oil categories

Purpose of test	Parameter
Quality assessment	• Acidity • Peroxide value • Halogenated solvents • Organoleptic properties—panel test • Absorbency in UV
Authenticity assessment	• Fatty acid composition • *Trans* fatty acids content • Sterol composition • Total sterols content • Erythrodiol and uvaol content • Waxes content • Difference between the actual and theoretical ECN42 triglyceride content • Stigmastadiene content • Saturated fatty acids at the 2-position (sum of palmitic and stearic acids as per cent total fatty acids)

Physicochemical and organoleptic criteria that allow the different grades of olive oil to be distinguished and authenticated have been developed and are laid down in Commission Regulation 2568/91 (European Communities, 1991), as most recently amended by Commission Regulation 455/2001 (European Communities, 2001a). These are summarized in Table 8.2. Most of the criteria are intended to ensure that the product and declared category are authentic. Acceptable ranges for fatty acids and sterols, for example, are included as the profiles for these components in different types of vegetable oil tend to be characteristic so are used to distinguish one oil from another. Stigmastadiene limits have also been established for checking authenticity. Stigmastadienes are breakdown products of the major sterol of vegetable oils, β-sitosterol, and are formed during edible oil refining. Elevated levels of stigmastadienes in 'extra virgin' or 'virgin' olive oils are, therefore, indicative of the presence of refined oil. Other criteria, such as acidity, peroxide value and organoleptic characteristics, are intended to give a measure of the quality of the product. The official methods of analysis that must be used to measure the various parameters are also laid down.

8.2.4 Origin labelling of olive oils

As mentioned previously, one of the main authenticity issues of concern within the olive oil sector is the possibility of false claims regarding the country or geographical region of origin. Commission Regulation 2815/98 (European Communities, 1998b) introduced optional provisions for designation of origin for the labelling of 'extra virgin' and 'virgin olive oil'. The designation of origin must relate to a geographical area that must be either a protected name or geographical origin (protected designation of origin—PDO, or protected geographical indication—PGI) under Council Regulation 2081/92 (European Communities, 1992b), and/or a Member State, the European Community (EC) or a third country. Regional designations may only be used for oils with PDO or PGI status and the area specified must be where the olives were grown and harvested. On the other hand, if the designation of origin indicates the EC or a Member State, it must correspond to the geographical area in which the 'extra virgin olive oil' or 'virgin olive oil' was extracted from the olives. In the case of blends, if more than 75% originates in the same Member State or in the Community, the main origin may be designated provided that it is followed by the indication 'selection of (extra) virgin olive oils more than 75% of which was obtained in (designation of origin)'. The provisions relate only to virgin grades and the Regulation specifically prohibits origin labelling of 'olive oil' and 'olive-pomace oil'. Provisions for enforcement in Great Britain of the Commission Regulation are contained in the *Olive Oil (Designation of Origin) Regulations 1999* (GB Statutory Instrument, 1999).

8.2.5 Review of olive oil classification and labelling

It is worth noting here that the classification and labelling of olive oils is currently being reviewed within the European Union. The background to this is given in the Commission's report to the Council and the European Parliament on its Quality Strategy for Olive Oil (Commission of the European Communities, 2000). This report highlights that the strategy was developed because the current classification system is confusing and that it no longer corresponds to the market situation. It was also considered that labelling rules are insufficiently precise and misleading to consumers. Further, there was a need to make improvements in the control of fraud. In particular, there is a great deal of concern about the dilution of lampante virgin oil with pomace-oil produced by centrifugation and also about the fraudulent addition of hazelnut oil to olive oil. In both cases, there is presently no robust analytical method available for detection.

8.2.5.1 Classification

Following the Commission's report and negotiations within the European Council, a revised classification system has been agreed and the new categories are specified in Council Regulation 1513/2001 (European Communities, 2001b). A number of important changes have been made.

- The description of the generic class 'virgin olive oils' has been amended to clarify that adjuvants that have a chemical or biochemical action (e.g. enzymes) may not be used during extraction to increase yields. The use of such adjuvants is considered by some to compromise the image of olive oil as a natural product. However, implicit in the revised definition is permission to use physical adjuvants, although this is acceptable only if the residue level is zero and use does not alter the composition of the oil. In this context, only food grade talc, as provided for in the Annex to Council Directive (EC) No. 2001/30 (European Communities, 2001c), may be used.

- Improvements by producers have led to an increase in production of 'extra virgin' and 'virgin' grades and a decrease in that of 'ordinary' and 'lampante' grades. To reflect this, the 'ordinary virgin olive oil' category has effectively been abolished by combining it with the 'lampante' category. Further, the maximum permitted acidity value (expressed as oleic acid) for 'extra virgin' oil has been reduced from 1.0 g per 100 g to 0.8 g per 100 g.
- The designation 'olive oil' has been expanded to 'olive oil: composed of refined olive oil and virgin olive oil'. This is much more meaningful and reflects more accurately exactly what this product is, i.e. a blend of refined olive oil and virgin olive oil. As a result of improvements in oil refining, it has also been possible to reduce the maximum acidity permitted for this grade of oil from 0.5 to 0.3 g per 100 g.
- The definition of 'crude olive-pomace oil' has been changed because this type of oil is now not only derived by solvent extraction of the olive pomace but also by physical extraction by means of a second centrifugation of the olive paste. The setting of limits for certain chemical criteria, aliphatic alcohols and wax, to differentiate lampante oil and second centrifugation oil, is currently under consideration.

The new definitions and categories of olive oil, which are listed in Table 8.3, come into effect on 1 November 2003, with the exception of that for 'crude olive-pomace oil', which is effective from 1 November 2001. This allows producers time to adjust to the new system.

8.2.5.2 Labelling

The Quality Strategy also covers labelling issues and, after discussions on the Commission's report, the Council concluded that further detailed consideration should be given to a number of issues (Council of the European Union, 2001). This includes the labelling of blends of olive oil with other vegetable oils. This was a difficult issue, particularly for the olive oil producer Member States (France, Greece, Italy, Portugal and Spain) that argue that the potential for adulteration and fraud is very high and that the image of olive oil is tarnished by such practices. Nevertheless, there appears to be no legal basis to ban such mixtures and there is no justification on consumer health grounds. Indeed, they are already on sale in some non-producer Member States, including the UK. It is accepted that there is a lack of analytical methods for verifying conclusively and

Table 8.3 Revised descriptions and definitions of olive oils and olive-pomace oils

Description	Definition
Virgin olive oils	This is a generic term and refers to oils from the fruit of the olive tree derived solely by mechanical or other physical means under conditions that do not lead to alteration of the oil, and which have undergone no treatment other than washing, decantation, centrifugation or filtration, but excluding oils obtained using solvents or *using adjuvants having a chemical or biochemical action*,[a] or by re-esterification processes and any mixtures with other oils.
Extra virgin olive oil	Virgin olive oil having a *maximum free acidity, expressed as oleic acid, of 0.8 g per 100 g* and the other characteristics laid down for this category.
Virgin olive oil	Virgin olive oil having a maximum free acidity, expressed as oleic acid, of 2.0 g per 100 g and the other characteristics laid down for this category.
Lampante virgin olive oil	Virgin olive oil having a *free acidity, expressed as oleic acid, of more than 2.0 g per 100 g* and the other characteristics laid down for this category.
Refined olive oil	Olive oil obtained by refining virgin olive oil, having a *maximum free acidity, expressed as oleic acid, of 0.3 g per 100 g* and the other characteristics laid down for this category.
Olive oil—composed of refined olive oils and virgin olive oils	Olive oil obtained by blending refined olive oil and virgin olive oil, other than lampante oil, having a *maximum free acidity, expressed as oleic acid, of 1.0 g per 100 g* and the other other characteristics laid down for this category.
Crude olive-pomace oil	Oil obtained by treating olive pomace with solvents *or by physical means or oil corresponding to lampante olive oil except for certain specified characteristics*, excluding oil obtained by means of re-esterification and mixtures with other types of oil, and having other the characteristics laid down for this category.
Refined olive-pomace oil	Oil obtained by refining crude olive-pomace oil, having a *maximum free acidity, expressed as oleic acid, of 0.3 g per 100 g* and the other characteristics laid down for this category.
Olive-pomace oil	Oil obtained by blending refined olive-pomace oil and virgin olive oil, other than lampante oil, having a *maximum free acidity, expressed as oleic acid, of 1 g per 100 g* and the other characteristics laid down for this category.

[a]Changes which have been introduced are italicized.

quantifying definitively the oils in a mixture but this is not sufficient justification for banning blends. General labelling rules apply and ensure that consumers are not misled as to the nature of these products and that compliance may be checked though audit trail.

With regard to geographical origin, there was agreement that a clear and unambiguous approach should be adopted and that the designation of origin should be based on the place of olive harvest and, if different, the place of pressing/processing. This is an improvement on the current rules as the organoleptic properties of the oil are dependent very much on the olives used and the conditions under which they were grown.

It has also been recognized that declaration of acidity of the oil may mislead consumers with respect to the quality because it is only meaningful within a single category of oil. Refining reduces the acidity, therefore, consumers could erroneously believe that 'olive oil' is of a higher quality than 'extra virgin olive oil'. In view of this, there is to be a prohibition on labelling of acidity unless additional information is given and on other indications falsely implying quality.

8.3 International standards—*Codex Alimentarius*

Internationally too, measures have been established to protect consumers and ensure the authenticity of the foods they purchase, namely the standards of the *Codex Alimentarius* ('food code'). These will be described in this section and in particular as they relate to edible fats and oils. However, in order to put this into context, some background information will be given first on the *Codex Alimentarius* organization and how it works.

8.3.1 Codex Alimentarius Commission

The *Codex Alimentarius* Commission (CAC) is an inter-governmental body sponsored jointly by the Food and Agricultural Organization (FAO) of the United Nations, and the World Health Organization (WHO). It was set up in 1962 and since then has developed a set of international standards, codes of practice, guidelines, etc. that encompasses all the principal food commodities. The aims of the *Codex Alimentarius* are to protect consumer health and to ensure fair practices in international food trade. Membership comprises governments of over 165 countries. International non-governmental organizations, including consumer groups and industry bodies, are also very much involved in the work of *Codex* and have observer status.

Much of the work of the CAC is delegated to various subsidiary committees. Part of the subsidiary structure is illustrated in Figure 8.1. Firstly, there are a number of regional committees which cover, Africa, Asia, Europe, Latin America and the Caribbean, and North America and the South West Pacific. These coordinate food standards activities within these regions. Secondly, there are a number of commodity committees that develop standards for defined groups of foodstuffs. This includes the Codex Committee on Fats and Oils (CCFO) for which the UK is the host country and for which the Food Standards Agency (FSA) provides the Secretariat. Finally, there a number of general committees which deal with horizontal issues affecting all foods. These include, for example, the Codex Committee on Food Additives and

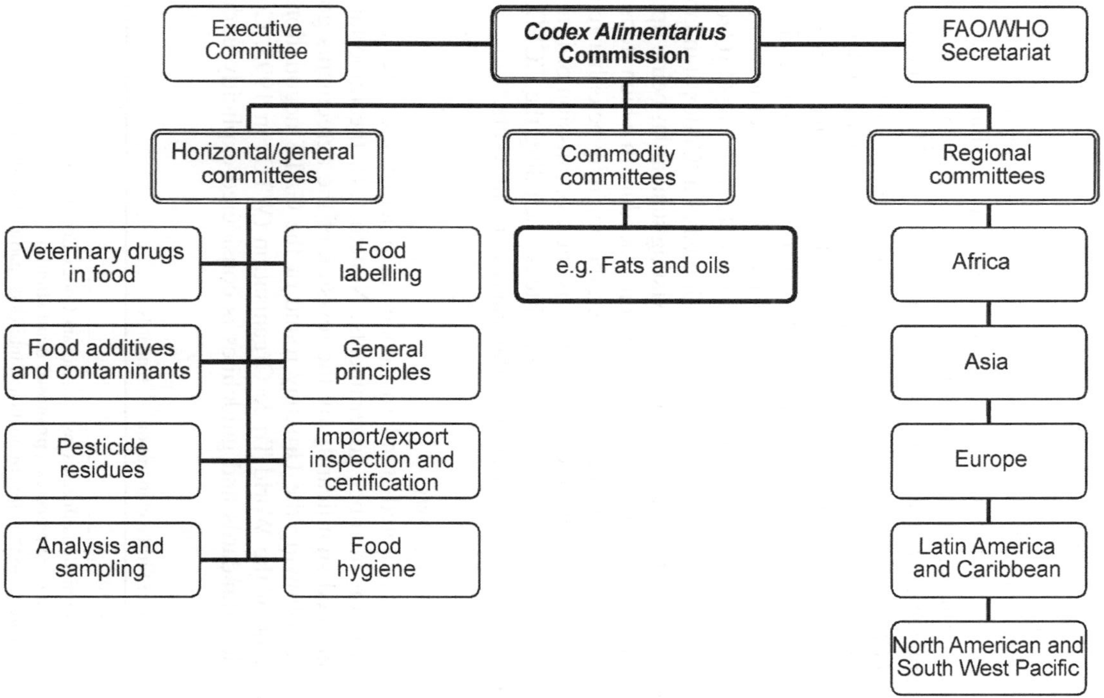

Figure 8.1 Structure of the *Codex Alimentarius* Commission.

Contaminants and the Codex Committee on Food Labelling. The activities of these committees will clearly have a bearing upon the development of standards for edible fats and oils by the CCFO.

8.3.1.1 Elaboration of Codex standards

The basis of Codex work is the elaboration of internationally recognized standards, guidelines and codes of practice. These are developed on the basis of the best scientific and technical information and advice available. The aim is to ensure that consumers may buy sound, wholesome food products free from adulteration and which are labelled and presented correctly. The standards, etc., are elaborated according to an eight-step procedure outlined in Table 8.4. The CAC first determines the need for a standard and assigns the work of developing it to the appropriate committee. The secretariat of the committee then arranges the preparation of a 'proposed draft standard'. The draft is subsequently circulated to governments and interested international organizations for comment at steps 3 and 6 and discussed by the Committee at steps 4 and 7. The CAC reviews the draft at steps 5 and 8 before it is formally adopted as a 'Codex standard'. In certain circumstances, for example where an existing standard is being updated, it is possible to follow an accelerated elaboration procedure in which steps 6 and 7 are omitted. In other circumstances, however, the text may be either retained at a certain step by the relevant committee or by the Commission or it may be sent back a step. As most commodity committees, including the CCFO, only meet biannually, the process of elaboration can take many years.

8.3.1.2 Status of Codex texts

Under Codex rules, adopted standards, codes of practice, etc., are not automatically binding on member governments. Nonetheless, the importance of Codex was enhanced significantly by the revision of the agreements governing world trade following the Uruguay round of the GATT negotiations and the establishment of the World Trade Organisation (WTO) in 1995. Conformity with Codex standards and guidelines is considered explicitly under the

Table 8.4 Codex procedure for the elaboration of standards

Step	Activity
1	Commission allocates task to relevant Codex Committee
2	Committee prepares a **proposed draft standard**
3	Circulation to Governments and interested parties for comment
4	Amendment in light of comments received and discussion at Committee session
5	Submission to Commission for formal adoption as a **draft standard**
6	Circulation to Governments and interested parties for further comments
7	Amendment in the light of comments and discussion at Committee session
8	Submission to Commission for formal adoption as a **Codex standard**

Sanitary and Phytosanitary (SPS) agreement (WTO, 2000a) and implicitly (by reference to international agreements) under the *Technical Barriers to Trade (TBT) agreement* (WTO, 2000b). As a consequence, if national food law is challenged as presenting an unjustified barrier to trade, under WTO's dispute settlement procedure, Codex standards and texts may be consulted and used as a key point of reference and guidance. This has given Codex texts a quasi legal status in international law. The full implications of this will, however, only become clearer as case law develops.

Further information on Codex and its activities may be found in '*Understanding the Codex Alimentarius*' (Joint FAO/WHO, 1999a) or on the Codex website (http://www.codexalimentarus.net).

8.3.2 Codex general labelling requirements

The *Codex General Standard for the Labelling of Pre-packaged Food* (Joint FAO/WHO, 1991a) is important with respect to authenticity of all foods, particularly the general principles which it establishes. First, that any pre-packaged food shall not be described or presented in a manner that is false, misleading or deceptive or is likely to create an erroneous impression regarding its character in any respect. Second, that any food shall not be described or presented by words, pictorial or other devices that refer to, or are suggestive either directly or indirectly, of any other product with which the food might be easily confused. Also important are the provisions for mandatory labelling which, as with national and European legislation, require that foods be marked with, *inter alia*, the 'name of the food', a list of ingredients, and the country of origin where its omission would mislead or deceive the consumer.

8.3.3 Codex standards for fats and oils

As well as the general labelling requirements, Codex has also elaborated a number of commodity standards for edible fats and oils through the CCFO. These standards contain identity and quality characteristics together with provisions relating to food additives, contaminants, hygiene and labelling and also specify methods of analysis. The provisions aim to ensure public health and safety and consumer protection as well as facilitating fair trade and preventing fraud. The work of the CCFO in developing the standards is ongoing but a major revision and simplification was initiated in 1991 following a recommendation by the CAC (Joint FAO/WHO, 1991b). Since then, group standards have been agreed for 'named animal fats' and 'named vegetable oils' and also far all other fats and oils (other than olive oils) that are not covered by these two standards (Joint FAO/WHO, 1999b,c,d). A separate standard is being retained for olive oils because of the detailed provisions needed for the different grades, but this is still under review (Joint FAO/WHO, 1993, 2001a).

8.3.3.1 *Standards for named animal fats and named vegetable oils*

These standards were adopted formally by the CAC in 1999 (Joint FAO/WHO, 1999e). The *Standard for Named Animal Fats* combines and updates provisions for lard, rendered pork fat, premier jus and edible tallow. The *Named Vegetable Oils Standard* covers 15 different oils of particular importance in international trade and also the palm oil fractions, palm olein and palm stearin (see Table 8.5).

In revising the provisions for vegetable oils, the CCFO have discussed widening the scope of the standard to include oils from new varieties of plants, produced either by traditional plant breeding techniques or through genetic engineering (Joint FAO/WHO, 1997, 1999f, 2001a). The Committee recognizes that this is a fast developing area. In view of this, it has agreed that in reaching a decision on inclusion of any new oils, consideration should be given to: the level of international trade; the justification for the inclusion of the oil or fat within

Table 8.5 Animal fats and vegetable oils covered by Codex Group Standards

Animal fats

- Pure rendered lard
- Lard subject to processing
- Rendered pork fat
- Rendered pork fat subject to processing
- Premier jus (oleo stock)
- Edible tallow (dripping)
- Edible tallow subject to processing

Vegetable oils (synonyms in brackets)

- Arachis oil (peanut oil, groundnut oil)
- Babassu oil
- Coconut oil
- Cottonseed oil
- Grapeseed oil
- Maize oil (corn oil)
- Mustardseed oil
- Palm kernel oil
- Palm oil
- Palm olein
- Palm stearin
- Rapeseed oil (turnip rape oil, colza oil, ravison oil, sarson oil, toria oil)
- Rapeseed oil—low erucic acid (low erucic acid turnip rape oil; low erucic acid colza oil, canola oil)
- Safflowerseed oil (safflower oil, carthamus oil, kurdee oil)
- Safflowerseed oil—high oleic acid (high oleic acid safflower oil, high oleic acid carthamus oil, high oleic acid kurdee oil)
- Sesameseed oil (sesame oil, gingelly oil, benne oil, ben oil, till oil, tillie oil)
- Soyabean oil (soybean oil)
- Sunflowerseed oil (sunflower oil)
- Sunflowerseed oil—high oleic acid (high oleic acid sunflower oil)
- Olive oil, virgin and refined, and refined olive-pomace oil

the scope of the standard; taxonomic information; and, where appropriate, the extent of the difference between the new oil and those already included in the standard. The first 'new' oils to be added following such consideration are 'high oleic acid sunflower oil' and 'high oleic acid safflower oil'. The standard has been amended accordingly and the new provisions adopted (Joint FAO/WHO, 2001b). Plans to add provisions for 'mid-oleic acid sunflower oil' and 'super palm olein' have also now been agreed (Joint FAO/WHO, 2001a).

The fats and oils standards aim to provide protection in terms of authenticity of the products covered and contain descriptions of each. They also recognize that the fatty acid composition is one of the main ways of identifying and distinguishing one oil/fat from another. Fatty acid specifications for each oil/fat are included as 'essential composition and quality factors' and are reproduced in Tables 8.6 and 8.7. For vegetable oils, it is also recognized that other compositional factors may be used to assist in assessing authenticity—iodine

Table 8.6 Codex fatty acid specifications for animal fats (% total fatty acids)

Fatty acid	Lard	Rendered pork fat	Premier jus	Tallow
C6:0	)	)	)	)
C8:0	)	)	)	)
C10:0	)<0.5 in total	)<0.5 in total	)<0.5 in total	)<0.5 in total
C12:0	)	)	)	)
C14:0	1.0–2.5	1.0–2.5	2–6	2–6
C14:ISO	<0.1	<0.1	<0.3	<0.3
C14:1	<0.2	<0.2	0.5–1.5	0.5–1.5
C15:0	<0.2	<0.2	0.2–1.0	0.2–1.0
C15:ISO	<0.1	<0.1	)<1.5 in total	)<1.5 in total
C15:ANTI ISO	<0.1	<0.1	)	)
C16:0	20–30	20–30	20–30	20–30
C16:1	2.0–4.0	2.0–4.0	1–5	1–5
C16:ISO	<0.1	<0.1	<0.5	<0.5
C16:2	<0.1	<0.1	<1.0	<1.0
C17:0	<1	<1	0.5–2.0	0.5–2.0
C17:1	<1	<1	<1.0	<1.0
C17:ISO	<0.1	<0.1	)<1.5 in total	)<1.5 in total
C17:ANTI ISO	<0.1	<0.1	)	)
C18:0	8–22	8–22	15–30	15–30
C18:1	35–55	35–55	30–45	30–45
C18:2	4–12	4–12	1–6	1–6
C18:3	<1.5	<1.5	<1.5	<1.5
C20:0	<1.0	<1.0	<0.5	<0.5
C20:1	<1.5	<1.5	<0.5	<0.5
C20:2	<1.0	<1.0	<0.1	<0.1
C20:4	<1.0	<1.0	<0.5	<0.5
C22:0	<0.1	<0.1	<0.1	<0.1
C22:1	<0.5	<0.5	not detected	not detected

Reproduced by kind permission of the Food and Agriculture Organization of the United Nations from: *Codex Standard for Named Animal Fats* (Codex Stan 211-1999).

Table 8.7 Codex fatty acid specifications for vegetable oils (% total fatty acids)

Fatty acid	Arachis oil	Babassu oil	Coconut oil	Cottonseed oil	Grapeseed oil	Maize oil	Mustardseed oil	Palm oil	Palm kernel oil	Palm olein
C6:0	ND	ND	ND–0.7	ND	ND	ND	ND	ND	ND–0.8	ND
C8:0	ND	2.6–7.3	4.6–10.0	ND	ND	ND	ND	ND	2.4–6.2	ND
C10:0	ND	1.2–7.6	5.0–8.0	ND	ND	ND	ND	ND	2.6–5.0	ND
C12:0	ND–0.1	40.0–55.0	45.1–53.2	ND–0.2	ND	ND–0.3	ND	ND–0.5	45.0–55.0	0.1–0.5
C14:0	ND–0.1	11.0–27.0	16.8–21.0	0.6–1.0	ND–0.3	ND–0.3	ND–1.0	0.5–2.0	14.0–18.0	0.5–1.5
C16:0	8.0–14.0	5.2–11.0	7.5–10.2	21.4–26.4	5.5–11.0	8.6–16.5	0.5–4.5	39.3–47.5	6.5–10.0	38.0–43.5
C16:1	ND–0.2	ND	ND	ND–1.2	ND–1.2	ND–0.5	ND–0.5	ND–0.6	ND–0.2	ND–0.6
C17:0	ND–0.1	ND	ND	ND–0.1	ND–0.2	ND–0.1	ND	ND–0.2	ND	ND–0.2
C17:1	ND–0.1	ND	ND	ND–0.1	ND–0.1	ND–0.1	ND	ND	ND	ND–0.1
C18:0	1.0–4.5	1.8–7.4	2.0–4.0	2.1–3.3	3.0–6.5	ND–3.3	0.5–2.0	3.5– 6.0	1.0–3.0	3.5–5.0
C18:1	35.0–69	9.0–20.0	5.0–10.0	14.7–21.7	12.0–28.0	20.0–42.2	8.0–23.0	36.0–44.0	12.0–19.0	39.8–46.0
C18:2	12.0–43.0	1.4–6.6	1.0–2.5	46.7–58.2	58.0–78.0	34.0–65.6	10.0–24.0	9.0–12.0	1.0–3.5	10.0–13.5
C18:3	ND–0.3	ND	ND–0.2	ND–0.4	ND–1.0	ND–2.0	6.0–18.0	ND–0.5	ND–0.2	ND–0.6
C20:0	1.0–2.0	ND	ND–0.2	0.2–0.5	ND–1.0	0.3–1.0	ND–1.5	ND–1.0	ND–0.2	ND–0.6
C20:1	0.7–1.7	ND	ND–0.2	ND–0.1	ND–0.3	0.2–0.6	5.0–13.0	ND–0.4	ND–0.2	ND–0.4
C20:2	ND	ND	ND	ND–0.1	ND	ND–0.1	ND–1.0	ND	ND	ND
C22:0	1.5–4.5	ND	ND	ND–0.6	ND–0.5	ND–0.5	0.2–2.5	ND–0.2	ND–0.2	ND–0.2
C22:1	ND–0.3	ND	ND	ND–0.3	ND–0.3	ND–0.3	22.0–50.0	ND	ND	ND
C22:2	ND	ND	ND	ND–0.1	ND	ND	ND–1.0	ND	ND	ND
C24:0	0.5–2.5	ND	ND	ND–0.1	ND–0.4	ND–0.5	ND–0.5	ND	ND	ND
C24:1	ND–0.3	ND	ND	ND	ND	ND	0.5–2.5	ND	ND	ND

Table 8.7 (continued)

Fatty acid	Palm stearin	Rapeseed oil	Rapeseed oil (low erucic acid)	Safflowerseed oil	Safflowerseed oil (high oleic acid)	Sesameseed oil	Soyabean oil	Sunflowerseed oil	Sunflowerseed oil (high oleic acid)
C6:0	ND	ND	ND	ND	ND	ND	ND	ND	ND
C8:0	ND	ND	ND	ND	ND	ND	ND	ND	ND
C10:0	ND	ND	ND	ND	ND	ND	ND	ND	ND
C12:0	0.1–0.5	ND	ND	ND	ND–0.2	ND	ND–0.1	ND–0.1	ND
C14:0	1.0–2.0	ND–0.2	ND–0.2	ND–0.2	ND–0.2	ND–0.1	ND–0.2	ND–0.2	ND–0.1
C16:0	48.0–74.0	1.5–6.0	2.5–7.0	5.3–8.0	3.6–6.0	7.9–12.0	8.0–13.5	5.0–7.6	2.6–5.0
C16:1	ND–0.2	ND–3.0	ND–0.6	ND–0.2	ND–0.2	0.1–0.2	ND–0.2	ND–0.3	ND–0.1
C17:0	ND–0.2	ND–0.1	ND–0.3	ND–0.1	ND–0.1	ND–0.2	ND–0.1	ND–0.2	ND–0.1
C17:1	ND–0.1	ND–0.1	ND–0.3	ND–0.1	ND–0.1	ND–0.1	ND–0.1	ND–0.1	ND–0.1
C18:0	3.9–6.0	0.5–3.1	0.8–3.0	1.9–2.9	1.5–2.4	4.8–6.1	2.0–5.4	2.7–6.5	2.9–6.2
C18:1	15.5–36.0	8.0–60.0	51.0–70.0	8.4–21.3	70.0–83.7	35.9–42.3	17–30	14.0–39.4	75–90.7
C18:2	3.0–10.0	11.0–23.0	15.0–30.0	67.8–83.2	9.0–19.9	41.5–47.9	48.0–59.0	48.3–74.0	2.1–17
C18:3	ND–0.5	5.0–13.0	5.0–14.0	ND–0.1	ND–1.2	0.3–0.4	4.5–11.0	ND–0.3	ND–0.3
C20:0	ND–1.0	ND–3.0	0.2–1.2	0.2–0.4	0.3–0.6	0.3–0.6	0.1–0.6	0.1–0.5	0.2–0.5
C20:1	ND–0.4	3.0–15.0	0.1–4.3	0.1–0.3	0.1–0.5	ND–0.3	ND–0.5	ND–0.3	0.1–0.5
C20:2	ND	ND–1.0	ND–0.1	ND	ND	ND	ND–0.1	ND	ND
C22:0	ND–0.2	ND–2.0	ND–0.6	ND–1.0	ND–0.4	ND–0.3	ND–0.7	0.3–1.5	0.5–1.6
C22:1	ND	>2.0–60.0	ND–2.0	ND–1.8	ND–0.3	ND	ND–0.3	ND–0.3	ND–0.3
C22:2	ND	ND–2.0	ND–0.1	ND	ND	ND	ND	ND–0.3	ND
C24: 0	ND	ND–2.0	ND–0.3	ND–0.2	ND–0.3	ND–0.3	ND–0.5	ND–0.5	ND–0.5
C24:1	ND	ND–3.0	ND–0.4	ND–0.2	ND–0.3	ND	ND	ND	ND

Reproduced by kind permission of the Food and Agriculture Organization of the United Nations from *Codex Standard for Named Vegetable Oils* (Codex Stan 210-1999).
ND, non detectable, defined as $\leq$0.05%.

value, desmethylsterols, tocopherols and tocotrienols—and specifications for these are also included. It is important that these specifications reflect fully the composition of fats and oils traded internationally. They must also take into account natural variations in composition, e.g. between oils derived from different varieties of a particular type of oilseed or with respect to the effects of geographical growing area and harvest year. The specifications were originally based on data supplied by the Leatherhead Food Research Association (King *et al.*, 1985; Turrell and Whitehead, 1989, 1990; Lee *et al.*, 1993) but the Committee keep the ranges under review and the standard has since been supplemented by information from other sources. Supplementary data is only accepted by the CCFO on the basis that it is derived from samples of commercially grown varieties of known authenticity and using internationally recognized standard methods of analysis (Joint FAO/WHO, 1999f).

For maize oil, an extra parameter has been introduced—the stable carbon isotope ratio (SCIR). Maize is one of a small number of plants that employs the Hatch–Slack or C4 pathway for photosynthesis rather than the more common Calvin cycle or C3 pathway. The C4 pathway is less discriminating against the stable ^{13}C isotope and, as a result, the ratio of ^{12}C to ^{13}C is significantly different to that of the other oils of major commercial importance. As a consequence, the SCIR may be used as a means of detecting the presence of other vegetable oils in maize oil. The Codex acceptable range is again based on data from the Leatherhead Food Research Association (Lee *et al.*, 1993; Rossell, 1994).

8.3.3.2 Draft standard for olive oils and olive-pomace oils

As mentioned earlier, a separate standard is being retained for olive oils and olive-pomace oils because of the number of different grades of these oils that must be covered and the detailed provisions that are needed for each. The draft text of the revised standard is based on the International Olive Oil Council's (IOOC) Trade Standard (IOOC, 1999). It has been considered at step 7 of the Codex procedure at the last three meetings of CCFO (Joint FAO/WHO 1997, 1999f, 2001a). At the 15th and 16th sessions in 1997 and 1999, discussions were deferred because the European Community and the IOOC were undertaking fundamental reviews of olive oil classification and standards such that it was not feasible to finalize the Codex text. Consequently, concerted efforts were made at the 17th session in 2001 to try and advance the draft but these proved to be fruitless. Much of the discussion concerned the linolenic acid (C18:3) content of olive oil. Some North African countries wanted to increase the permitted maximum from 0.9 to 1.0% as authentic oils from these regions contain naturally higher levels than those from European countries. The European Commission, however, was unable to accept this and it was not possible to achieve consensus. The Committee recognized that it would not, therefore, be possible to finalize the text at this stage so returned it to step 6 for a further round of Government comments and discussion at the next CCFO session.

One important consequence of the Committee's failure to make progress is that the existing Standard (Joint FAO/WHO, 1993) remains in place and will be the reference used by WTO in any trade dispute. This text is very out of date and revised provisions are needed urgently in order to ensure consumer protection and to facilitate fair trade. It is thus important that the European Commission and the IOOC harmonize their rules and that agreement is reached on the Codex text when it is discussed again at the 18th CCFO session in 2003. In this respect, a commitment was made by the European Commission to the Council to work constructively with the IOOC in order to harmonize standards (Council of the European Union, 2001).

8.4 Enforcement and monitoring of labelling legislation

Enforcement of food law in the UK is the responsibility of local authorities and port health authorities. They may take prosecutions to uphold the legislation in place or may carry out surveys to monitor the food supply. The Food Standards Agency also has a research and development (R&D) and monitoring programme. This deals mainly with food safety issues but it also examines the nutritional adequacy of the diet as well as misdescription of foods. The programme aims to complement the work of the enforcement authorities and provide support by developing methodology and by highlighting problem issues.

8.4.1 The FSA food authenticity research and development programme

The aim of the Agency's R&D programme is to underpin policy. One of the policy aims is to maintain consumer choice by encouraging clear, accurate and informative labelling and reducing the likelihood of misdescribed food being sold on the UK market. The Authenticity R&D programme aims to develop reliable and cost-effective analytical techniques for assessing the authenticity of food especially as it relates to issues arising from compositional and labelling requirements. The methods may then be used either by the Agency to carry out surveys or, by enforcement, consumer and industry groups, to check that food is not misdescribed. The programme deals with issues across all the major food groups, including fats and oils, and also covers horizontal issues of interest to consumers such as the geographical origin of food, or whether food labelled as 'organic' is what it claims to be.

Currently, work in the fats and oils sector is focusing on developing methods for the detection of fraudulent addition of hazelnut oil to olive oil. As mentioned earlier, this is currently a major issue of concern because there are no reliable methods for its detection. The chemical composition of hazelnut oil is very similar to olive oil, at least in terms of the main components, so it is not possible to detect this type of adulteration using fatty acid or sterol analysis, etc. The work funded by the Agency has thus concentrated on finding potential markers for hazelnut oil from among the minor components. Preliminary studies at the University of Reading, UK, have identified such a marker among the non-volatile polar components (Gordon *et al.*, 2001). Further study is now underway to try to identify this compound definitively, establish its natural variability, and to explore its potential for detection of refined as well as unrefined oil. Another project was commissioned at the UK's Central Science Laboratory (CSL) which investigated the use of the volatile component filbertone (2-methyl heptenone), as a marker for hazelnut oil. Despite encouraging claims in the literature from other laboratories (Blanch and Jauch, 1998; Caja *et al.*, 1999; Pfnuer *et al.*, 1999), however, the results obtained raised a number of important questions (Brereton and Crews, 2001). These will be addressed in ongoing work at the CSL, which the Agency is part-funding, as part of the EU-collaborative project know as MEDEO (detailed information of which is available on the MEDEO website, http://www.cica.es/aliiens/igmedeo). A total of 14 laboratories are participating in this work which will be investigating the hazelnut oil problem using: methods based on the profile of volatile compounds; isotopic methods based on stable isotope mass spectrometry (SIRMS) and gas chromatography-pyrolysis-isotope ratio mass spectrometry (GC-Py-IRMS); spectrometric methods based on nuclear magnetic resonance (NMR), Fourier transform magnetic resonance imaging (FT-MIR) and Fourier transform Raman (FT-Raman) and chromatographic methods based on triacylglycerides and sterol ester composition. A further project, assessing the use of molecular markers, is also underway at the Campden and Chorleywood Research Association, UK.

8.4.2 The FSA food authenticity surveillance programme

Under this programme, surveys are carried out to investigate whether authenticity problems exist at a given point in time and, if so, to give an indication of the extent of the problem. They also provide an opportunity to test newly developed methods emerging from the food authenticity R&D programme and check their suitability for enforcement purposes.

The surveillance programme is overseen by the Agency's Working Party on Food Authenticity (WPFA). Membership of the Working Party comprises officials, together with representatives from consumer organizations, enforcement bodies, academia, the retail and hospitality sectors, and the food industry. The WPFA helps the Agency prioritize issues that need investigation, evaluate methodology to be used and agrees protocols to carry out surveys. The findings of all the surveys undertaken by the WPFA since its inception in 1992 (under the auspices of the then UK Ministry of Agriculture, Fisheries and Food) have been made public. However, in line with the Agency's objectives to put the consumer first and to be open and accessible, the results of analyses are now given together with full information on brand owners or suppliers. Where any problem results occur, the individual companies involved are made aware of the findings relating to their products and are asked for their comments, which are also published. The relevant enforcement bodies are also informed so that they can decide if any follow-up action is necessary. In this way the authenticity programme highlights areas where increased monitoring would seem to be appropriate and, thereby, enables enforcement officers and industry to focus their efforts and resources.

8.4.2.1 Vegetable oil surveys

Within the fats and oils sector, the results of two surveys have so far been published. The first dates back to 1995 when the accuracy of the description of the oil type specified on the label of maize (corn), sunflower, palm and groundnut oils sold through retail and wholesale outlets in the UK was investigated (MAFF, 1995; Stewart and Gillatt, 1996). The particular oils chosen were those that, at that time, traded at a premium and were thus likely candidates for adulteration. A total of 290 samples were collected and were screened on the basis of the fatty acid composition for the presence of oils other than that declared on the label. The SCIR of maize oils and the slip melting point of palm oils (as a means of detecting the presence of palm stearin) were also measured. The authenticity of survey samples was assessed by comparing their chemical composition and SCIR with that of the relevant pure oil. Samples which were found not to comply with the purity criteria for the oil named on the product label were further analysed to determine the desmethysterol and tocopherol and tocotrienol (tocol) composition to try to establish the identity of the undeclared oil. Rapeseed and soyabean oil are the likely adulterants and are characterized by high concentrations of brassicasterol and δ-tocopherol, respectively, so elevated levels of these components in other oils are indicative of the presence of these cheaper oils.

The majority of samples (81%) were considered to be correctly labelled with respect to the named oil. Approximately 11% contained between 3% and 5% undeclared oil, which was considered to be higher than would be expected if good manufacturing practice had been followed. A further 7% of samples were found to contain in excess of 5% undeclared oil which was considered to indicate deliberate adulteration. Form the desmethylsterol/tocol analysis, most adulterated samples were believed to contain rapeseed oil. The retailers and producers of these oils were informed of the results so that manufacturing practices could be reviewed and improved. The results for maize oils, in particular, were of concern as 35% of these contained greater than 3% undeclared oil. In view of this, the Agency is currently undertaking a repeat survey of maize oil to check whether or not manufacturers have improved their practices. The results of this survey were due to be published in Spring 2002.

A survey of olive oils was carried out in 1998, again to assess the accuracy of the description of the oil type specified on the label. A total of 125 samples were collected from retail outlets in the UK and included products described as 'extra virgin olive oil', 'virgin olive oil' and 'olive oil'. The addition of undeclared oils, such as lower grades of olive oil or other types of vegetable oil (e.g. rapeseed or soyabean oil), was investigated using a number of the chemical parameters listed in Commission Regulation 2568/91 (European Communities, 1991). These included stigmastadiene analysis for the detection of refined oils in virgin oils, fatty acid analysis and equivalent carbon number (ECN42) determination for the detection of other vegetable oils, and wax content for the detection of pomace oils.

The results of analysis were compared with the limits specified in the Commission Regulation (Joint MAFF and Department of Health Food Safety and Standards Group, 1999a,b). For the purposes of the survey, and in line with the usual practice followed by UK enforcement authorities, where samples were found to lie outside the statutory limits, an additional tolerance was added to take account of potential variations in the analytical methodology. The Commission Regulation makes it clear that an oil may not be considered to fit its description if any one of its characteristics lies outside the limits laid down. On this basis, the majority of the oils analysed were found to be correctly described and only four samples were found to exceed European Commission limits for one or more chemical criteria used to distinguish and authenticate the different grades of olive oil.

8.5 Conclusions

Consumers need meaningful and honest information to enable them to make an informed choice about their diet and the foods they purchase. Protection against mislabelling and misdescription through legislation is an important part of food control. To ensure that this legislation is effective, enforcement and monitoring of compliance are essential. Review and revision is needed too, particularly at the international level, in order to keep apace with changing and developing technologies and product innovations. Research to develop new analytical methods is likely to be an ongoing but ever more difficult task as those who seek to gain financially find increasingly sophisticated ways of adulterating food.

References

Blanch, G.P. and Jauch, J. (1998) Enantiomeric composition of filbertone in hazelnuts in relation to extraction conditions. Multidimensional gas chromatography and gas chromatography mass spectrometry in the single ion monitoring mode of a natural sample. *J. Agric. Food Chem.*, **46**, 4283–4286.

Brereton, P. and Crews, C. (2001) *Detection of Hazelnut Oil Addition to Extra Virgin Olive Oil. Final Report to the Food Standards Agency*. Central Science Laboratory, York, UK.

Caja, M.M., Ruiz del Castillo, M.L., Herraiz, M. and Blanch, G.P. (1999) Study of the enantiomeric composition of chiral constituents in edible oils by simultaneous distillation–extraction. Detection of adulterated olive oils. *J. Am. Oil Chem. Soc.*, **76**, 1027–1030.

Commission of the European Communities (2000) *Commission Report to the Council and the European Parliament on the Quality Strategy for Olive Oil.* COM(2000) 855 final CNS 2000/0358.

Council of the European Union (2001) 2396th Council meeting—Agriculture, Brussels, 23 July, 2001. Press Release 10928/01 (presse 286). Council of the European Union.

European Communities (1966) Council Regulation (EEC) 136/66 on the common organisation of the market in oils and fats. *Official Journal of the European Communities*, **No. 172**, 30.9.1966, p. 3025.

European Communities (1978) Council Directive 78/112/EEC on the approximation of the laws of the Member States relating to the labelling, presentation and advertising of foodstuffs for sale to the ultimate consumer. *Official Journal of the European Communities*, **L33**, 8.2.1979, p. 1.

European Communities (1991) Commission Regulation (EEC) No 2568/91 on the characteristics of olive oil and olive-residue oil and on the relevant methods of analysis. *Official Journal of the European Communities*, **L248**, 5.9.1991, p. 1.

European Communities (1992a) Council Regulation (EEC) No 356/92 amending Regulation No 136/66/EEC on the establishment of a common organisation of the market in oils and fats. *Official Journal of the European Communities*, **L39**, 15.2.1991, p. 1.

European Communities (1992b) Council Regulation (EEC) No 2081/92 on the protection of geographical indications and designations of origin for agricultural products and foodstuffs. *Official Journal of the European Communities*, **L208**, 24.7.1992, p. 1.

European Communities (1998a) Council Regulation (EEC) No 1638/98 amending Regulation No. 136/66/EEC on the establishment of a common organisation of the market in oils and fats. *Official Journal of the European Communities*, **L210**, 28.7.98, p. 32.

European Communities (1998b) Commission Regulation (EC) No 2815/98 concerning marketing standards for olive oil. *Official Journal of the European Communities*, **L349**, 24.12.98, p. 8.

European Communities (2000) European Parliament and Council Directive 2000/13/EEC on the approximation of the laws of the Member States relating to the labelling, presentation and advertising of foodstuffs. *Official Journal of the European Communities*, **L109**, 6.5.2000, p. 29.

European Communities (2001a) Commission Regulation (EC) No 455/2001 amending Regulation 2568/91 on the characteristics of olive oil and olive-residue oil and on the relevant methods of analysis. *Official Journal of the European Communities*, **L65**, 7.3.2001, p. 9.

European Communities (2001b) Council Regulation (EC) No 1513/2001 amending Regulation No. 136/66/EEC on the establishment of a common organisation of the market in oils and fats. *Official Journal of the European Communities*, **L201**, 26.7.2001, p. 4.

European Communities (2001c) Commission Directive 2001/30/EC amending Directive 96/77/EC laying down specific purity criteria on food additives other than colours and sweeteners (Text with EEA relevance). *Official Journal of the European Communities*, **L146**, 31.5.2001, p. 1.

GB Statutory Instrument (1996) *Food Labelling Regulations 1996.* SI No. 1499. London: HMSO.

GB Statutory Instrument (1987) *Olive Oil (Marketing Standards) Regulations 1987.* SI No. 1783. London: HMSO.

GB Statutory Instrument (1999) *Olive Oil (Designation of Origin) Regulations 1999.* SI No. 1513. UK: The Stationery Office Ltd.

Gordon, M.H., Covell, C. and Kirsch, N. (2001) Detection of pressed hazelnut oil in admixtures with virgin olive oil by analysis of polar components. *J. Am. Oil Chem. Soc.*, **78**, 621–624.

IOOC (International Olive Oil Council) (1999) *Trade Standard Applying to Olive Oil and Olive-pomace Oil.* COI/T.15/NC no. 2/Rev. 9, 10 June, 1999.

Joint FAO/WHO Food Standards Programme (1991a) *Codex General Standard for the Labelling of Pre-packaged Foods.* Codex Stan 1-1985 (Rev. 1-1991).

Joint FAO/WHO Food Standards Programme (1991b) *Codex Alimentarius Commission: Report of the Nineteenth Session, Rome, 1–10 July, 1991.* Food and Agricultural Organization of the United Nations and World Health Organization, Rome, 1991.

Joint FAO/WHO Food Standards Programme (1993) Codex Standard for olive oil, virgin and refined, and refined olive-pomace oil. Codex Stan 33-1981. In *Codex Alimentarius: Fats, Oils and Related Products*. vol. 8. Food and Agricultural Organization of the United Nations and World Health Organization, Rome, 1993.

Joint FAO/WHO Food Standards Programme (1997) *Report of the Fifteenth Session of the Codex Committee on Fats and Oils, London, 4–8 November 1997*. Codex Alimentarius Commission, Alinorm 97/17.

Joint FAO/WHO (1999a) *Understanding the Codex Alimentarius*. Food and Agriculture Organization of the United Nations and World Health Organization, Rome, 1999.

Joint FAO/WHO Food Standards Programme (1999b) *Codex Standard for Named Animal Fats*. Codex Stan 211–1999.

Joint FAO/WHO Food Standards Programme (1999c) *Codex Standard for Named Vegetable Oils*. Codex Stan 210–1999.

Joint FAO/WHO Food Standards Programme (1999d) *Codex Standard for Fats and Oils not Covered by Individual Standards*. Codex Stan 19-1981 (rev. 2-1999).

Joint FAO/WHO Food Standards Programme (1999e) *Codex Alimentarius Commission: Report of the Twenty-first Session, Rome, 28 June–3 July, 1999*. Food and Agricultural Organization of the United Nations and World Health Organization, Rome, 1999.

Joint FAO/WHO Food Standards Programme (1999f) *Report of the Sixteenth Session of the Codex Committee on Fats and Oils, London, 4–8 March 1999*. Codex Alimentarius Commission, Alinorm 99/17.

Joint FAO/WHO Food Standards Programme (2001a) *Report of the Seventeenth Session of the Codex Committee on Fats and Oils, London, United Kingdom, 19–23 February 2001*. Codex Alimentarius Commission, Alinorm 01/17.

Joint FAO/WHO Food Standards Programme (2001b) *Codex Alimentarius Commission: Report of the Twenty-fourth Session, Geneva, 2000*. Food and Agricultural Organization of the United Nations and World Health Organization, Rome, 2001.

Joint Food Safety and Standards Group (of the Ministry of Agriculture, Fisheries and Food and the Department of Health) (1999) Authenticity of olive oils. *Food Surveillance Information Sheet No. 180*. Food Standards Agency, UK.

King, B., Sibley, I. and Zilka, S.A. (1985) Authenticity of edible vegetable oils and fats, Part VII: maize oil. *Leatherhead Food RA Research Reports No. 522*. Leatherhead Food RA, UK.

Lee, K., Gillatt, P. and Rossell, J.B. (1993) Authenticity of edible vegetable oils and fats, Part XX: determination of maize oil purity by stable carbon isotope ratio analysis (SCIRA). *Leatherhead Food RA Research Reports No. 719*. Leatherhead Food RA, UK.

MAFF (Ministry of Agriculture, Fisheries and Food) (1995) MAFF single seed vegetable oil surveillance exercise. *Food Surveillance Information Sheet No. 77*. Food Standards Agency, UK.

MAFF (Ministry of Agriculture, Fisheries and Food) (1999) *Guidance Notes on Nutrition Labelling*. Food Standards Agency, UK.

Pfnuer, P., Matsui, T., Grosch, W., Guth, H., Hofmann, T. and Schieberle, P. (1999) Development of a stable isotope dilution assay for the quantification of 5-methyl-(*E*)-2-hepten-4-one: Application to hazelnut oils and hazelnuts. *J. Agric. Food Chem.*, **47**, 2044–2047.

Rossell, J.B. (1994) Stable carbon isotope ratios in establishing maize oil purity. *Fett Wissenschaft Technologie*, **96**, 304–308.

Stewart, C.A. and Gillatt, P.N. (1996) Authenticity of single seed vegetable oils—a survey of the UK market. *J. Assoc. Public Analysts*, **32**, 179–214.

Turrell, J.A. and Whitehead, P.A. (1989) Authenticity of edible vegetable oils and fats, Part XVIII: analysis of additional samples of sunflower-seed, groundnut and maize germ oils. *Leatherhead Food RA Research Reports No. 637*. Leatherhead Food RA, UK.

Turrell, J.A. and Whitehead, P.A. (1990) Authenticity of edible vegetable oils and fats, Part XVI: analysis of additional samples of palm, soya-bean and rapeseed oils. *Leatherhead Food RA Research Reports No 665*. Leatherhead Food RA, UK.

UK Parliament (1968) *Trades Description Act 1968*, Ch. 19. London, HMSO.

UK Parliament (1990) *Food Safety Act 1990*, Ch. 16. London, HMSO.

World Trade Organisation (2000a) Agreement on the application of sanitary and phytosanitary measures, in *The Legal Texts—The Results of the Uruguay Round of Multilateral Trade Negotiations*. Cambridge University Press, pp. 59–120.

World Trade Organisation (2000b) Agreement on technical barriers to trade Agreement, in *The Legal Texts—The Results of the Uruguay Round of Multilateral Trade Negotiations*. Cambridge University Press, pp. 121–142.

Index

Printed and bound by CPI Group (UK) Ltd, Croydon, CR0 4YY

13/09/2026

14961030-0002